MULTIMEDIA AND THE INTERACTIVE DISPLAY

in museums, exhibitions and libraries

Signe Hoffos
Multimedia Ventures

Library and Information Research Report 87

BRITISH LIBRARY CATALOGUING IN PUBLICATION DATA
A CIP record for this book is available from the British Library
ISBN 0 7123 3268 5

This report is published by the British Library and distributed by the British Library Publications Sales Unit, Boston Spa, Wetherby, West Yorkshire LS23 7BQ, UK.

It is distributed in the USA by the American Library Association, 50 East Huron Street, Chicago, Illinois 60611.

In Japan, by Kinokuniya Co. Ltd., PO Box 55, Chitose, Tokyo 156.

ABSTRACT

Multimedia and the interactive display in museums, exhibitions and libraries is a two-part report which combines an introduction to multimedia technology with examples of its application. The first part addresses fundamental concepts and technical issues to explain how interactive multimedia systems work, and introduces platforms including LaserDisc and interactive video (IV), compact disc (CD-DA, CD-ROM/XA, CD-I, CDTV, Photo CD), Digital Video Interactive (DVI), holograms and high-definition television (HDTV). The second part describes over 30 projects at 26 sites, with first-hand observation, comments and advice. The report includes brief descriptions of many more applications and listings of sites worldwide, plus sources of further information.

ACKNOWLEDGEMENTS

The author would like to thank Susi Barrett of New Media and Piers Lea of Line TV for photographs; Hariet Phillips of the Design Museum, Tim Roderick of Norfolk TVEI and Florence Cusson of Opus Species for HyperCard and SuperCard images; and the Voyager Company of California for permission to reproduce screens from the *Ludwig van Beethoven: Symphony Nº 9* CD Companion.

THE AUTHOR

Signe Hoffos has been writing about interactive video and multimedia since 1982. She has contributed articles to many British and American magazines and anthologies, and spoken at events in Britain and Europe. She wrote *Interactive Video* (EPIC/Sigma 1983), *The CD-I Designers' Guide* (New Media/McGraw-Hill 1992) and edited several publications for the National Interactive Video Centre (NIVC), including the newsletter *Interactive Update, The Interactive Video Yearbook* and *Interactive Video Courseware Catalogue.* With Multimedia Ventures, she co-edited *The European Multimedia Yearbook 1992,* and provides consultancy and research services to the multimedia industry.

Printed from the author's disk by The College Hill Press Limited
37 Webber Street, London SE1 8QW, UK

CONTENTS

INTRODUCTION

The British Library is of course famous for its illuminated manuscripts, rare printed books, and records of historic events, great works and famous lives. However, with the move to its new building at St. Pancras in 1993, the British Library intends to complement the treasures of its historic collections with displays and information systems which will set a standard for modern exhibition design.

This report was originally commissioned by the British Library in the spring of 1990, as research for the initial planning of galleries and displays for the new building.

That report was written under the ægis of the UK National Interactive Video Centre (NIVC), for which the author edited publications including *The Interactive Video Yearbook* and the periodical *Interactive Update*. With the demise of the NIVC, late in 1990, the author and two colleagues formed Multimedia Ventures, a consultancy whose first project was *The European Multimedia Yearbook 1992*.

In the autumn of 1991, the British Library and Multimedia Ventures agreed to revise this report for publication to a wider readership. The introduction to multimedia and interactive technology was updated and enlarged, and the descriptions of work in museums, galleries and touring exhibitions were expanded to include a broader range of projects, as well as progress reports from installations first contacted some 18 months previously.

The technology, of course, will continue to change apace, just as the number and variety of interactive displays are bound to grow. The aim of this report is simply to give readers a grounding in the basic principles and vocabulary of the multimedia market, so that they might then pursue their own research more productively.

- *The New British Library* (pages 2-3) describes the terms under which the original report was commissioned, and suggests points which any institution might consider in researching and developing the use of new media.
- The *Management Summary* (pages 4-6) identifies key points and offers an overview specifically for senior decision-makers who may lack the time or inclination to study the entire report.
- *Interactivity and Multimedia* (pages 7-14) and *Delivery Systems* (pages 15-22) address technical issues with an introduction to basic concepts and essential hardware typical of any interactive multimedia presentation.
- *Multimedia Platforms* (pages 23-37) describes various digital, optical and audio-visual media now available to exhibition designers.
- The projects (pages 38-75) include both permanent and temporary installations with features of interest. Most are in Britain, simply because these descriptions were written from observation and interviews with designers and their clients, within a timescale and budget which did not encourage foreign travel.
- The balance of the report comprises some further comments from programme-makers, a few necessarily general remarks about the potential market for publishing on disc, and sources of further information and contacts. This includes lists of known projects in Europe and North America, and addresses for institutions and suppliers who contributed to this report.

With two rounds of research in two years, this report necessarily combines observations made in the summer of 1990 with information collected virtually up to press day, in April 1992. Some of those who contributed have since changed jobs, so the opinions quoted here, however valid, do not always reflect the corporate views of the institutions themselves. Equally, while over 50 organizations were contacted during the two periods of research, those who were most forthcoming with interviews and information are, not unnaturally, the best represented.

No matter how this market develops, the decision to use interactive multimedia systems is undoubtedly sound, and a foot on the learning curve now will make the eventual ascent into newer technologies easier; however, the initial choice of methods and media should be most carefully considered.

This report aims to provide information which decision-makers need to evaluate the many products and services now available to them. The appendices suggest further sources for those who wish to learn more by desk research and active investigation.

There are some serious decisions to make here, and no 'right answer'. This is one of many fields where any number of experts may speak with equal authority but hold quite different views. The pages which follow aim to bring all those who may be involved in the decision-making process to a common level of understanding, and to offer some unbiased information on the choices available.

Signe Hoffos, London, April 1992

THE NEW BRITISH LIBRARY

The British Library was created in 1973 from three national institutions – the British Museum Library, the National Central Library, and the National Lending Library for Science and Technology – whose reading rooms, offices and collections occupy several sites in central and north London, and West Yorkshire.

In 1993, the British Library will move to a new 12-storey building which covers nearly 20 acres of ground near St. Pancras Station in the Euston Road. The move will allow the British Library to bring together many of its scattered collections, staff and services, and provide both storage space and working facilities designed to serve one of the world's largest and greatest national libraries.

THE COLLECTIONS

Exhibitions in the British Library's new building, as now, will be a combination of permanent displays, 'major' temporary exhibitions and smaller changing exhibitions.

Although many visitors think of a library's treasures in terms of illuminated manuscripts and rare printed books, the full scope of its collections range from literary autographs and the documents of the India Office to market research reports and British and foreign patents. This includes a great deal of scientific, industrial and commercial material, together with work of artistic and literary value, and ephemera of many kinds.

The collections of the Map Library, the Music Library, the Philatelic Collection and the Science Reference Information Service extend the range of potential exhibition material well beyond the familiar world of books and manuscripts. Furthermore, although the National Sound Archive and the Newspaper Library will not move to the new building, their respective collections add another dimension to the variety of material which will be displayed and studied in the Library's new exhibition halls.

The British Library also has both objects and collections on themes of considerable general and academic interest, from children's books to the holy books of many religions. Individual items such as Magna Carta, the Lindisfarne Gospels, the Gutenberg Bible and the Diamond Sutra attract people with no particular interest in other parts of the collection, and certain of these warrant accessible permanent displays of their own.

The Library also has many examples of material illustrative of other general topics. The permanent exhibition will include information on the history of writing, of publishing and printing, and of audio recording.

The British Library is also, of course, one of the world's most important and active research libraries, handling some 25,000 separate book movements in any working day. The move to a new purpose-built site is certainly the occasion to draw attention to the scale and scope of this activity, and to help promote some of the Library's many services to the public.

Permanent and changing exhibitions are governed by the Library's obligation, as one of the nation's great storehouses, to display (and protect) its treasures and, in some cases, by the terms of bequests and loans through which objects were donated to its collections.

In designing and executing its exhibitions, the British Library can draw on both in-house resources and external contractors. There is already a large collection of still images, recorded mainly on microfilm and Ektachrome, which are used as reference and for reproduction in other media. There is an in-house video unit, with broadcast facilities, which will supply some production services when the time comes to produce source material for the projected multimedia display systems.

THE NEW EXHIBITION GALLERIES

The new building will afford 1350m^2 of dedicated exhibition space on two levels in adjacent areas. There is also a scattering of potential sites for small displays in the entrance hall and public rooms of the new Library.

This represents less actual floorspace than is now occupied within the British Museum, but amply compensates for the loss of square metres by providing the Library with exhibition facilities which are specifically designed to meet both the environmental and creative demands of modern display techniques. For all their elegance, the present galleries pose many limitations on the space actually available and the ways in which it can be used – there are, for example, no walls on which prints or paintings can be hung, and power sources for lighting and other effects are limited.

The main exhibition area lies just beyond the public entrance to the new building, which leads from a forecourt on the Euston Road. The 'treasures' of the British Library – objects such as the

Lindisfarne Gospels and the Diamond Sutra – will be displayed in permanent exhibitions on the upper floor of the new galleries. The cultural and material value of these exhibits places considerable demands on the design of this area, which must incorporate a high level of security and meticulous environmental control without making the objects themselves inaccessible. Low-level fibre optic lighting will help to address the familiar dilemma of illuminating objects which are vulnerable to damage from light.

However, these traditional objects of veneration form only part of the exhibition as a whole, and this would be a natural place to complement ancient arts with modern information systems. The history of printing and publishing spans a wide range of crafts and skills, from paper-making and fine binding to automation, and multimedia systems afford a new and imaginative way of presenting a wealth of information and images from archival material to on-line news services. Displays of this kind will be combined with more traditional forms of interactivity in the smaller of the two ground floor galleries.

The second of the ground floor exhibition galleries will house a variety of changing exhibitions which, as now, will combine 'major' events – which last perhaps four or six months, and attract international attention from scholars and tourists alike – with smaller exhibitions on topical themes or items of general interest.

The variation in ceiling heights in this area poses a combination of opportunities and challenges. Visitors first pass through a relatively low and narrow entrance foyer which leads to a large, light hall with ceilings which sweep to the very top of the exhibition area. Cherry-pickers will be needed even to change ceiling lights in this area, and to mount and remove banners, boards and other large suspended display material.

The upper exhibition gallery overlooks this large hall, and there is considerable potential for displays which create visual and conceptual links between the two spaces and their respective exhibits. Equally, it may sometimes be desirable to separate the upper and lower halls through appropriate display techniques. The ceilings should be equal to suspending all but the heaviest objects, so large passive display monitors could be suspended from the ceiling to complement smaller, active information terminals on the floor below. (Platform flooring and removable carpet tiles are to be used throughout this area, to facilitate installation and maintenance of both permanent and temporary displays.)

Some smaller spaces might also be used for interactive display systems elsewhere in the building. These would typically involve interactive point-of-sale/information (POS/POI) systems, and offer several opportunities for publishing on new media (particularly, CD); both these themes are discussed below.

ATTENDANCES

The move to the new building should make life easier for the Library's readers, who currently number around 350,000 a year, and represent every kind of academic and commercial activity from students to professional researchers.

However, the effect of the move on visitors of the other kind – who come as tourists or for special exhibitions and lectures – is less certain. Many people are undoubtedly now drawn to the British Library through the doors of the British Museum, which is one of the country's top attractions, with some 5,061,287 visits to its main site in Bloomsbury in 1991. The intimacy of the two exhibition areas makes it almost impossible at present to distinguish visits specifically to the British Library from visits to the British Museum as a whole. In contrast, the British Museum's own ethnography collection, the Museum of Mankind, in the same year attracted only 349,135 visits to its separate site in Mayfair.

By way of comparison, two relatively new and slightly offbeat London attractions, the Bank of England Museum in the City (which is free) and the Design Museum in Butlers Wharf (which charges admission), both estimate that they attract over 100,000 visitors a year.

The new location is just off the beaten track, with few other attractions of interest in the area, but is still easily accessible by all forms of public transport. The new building will be easy to find: persuading people to enter for the first time may be the challenge here as with many monumental public buildings. Attractive and friendly interactive information systems may be one key to establishing the British Library as an interesting and rewarding place to visit, for Londoners and occasional visitors to the capital as well as tourists in search of Shakespeare and Magna Carta.

MANAGEMENT SUMMARY

The terms interactive video (IV) and multimedia currently describe systems which, in material terms, combine the processing power of a computer with audio-visual (AV) resources including still and moving pictures, and sound.

- Interactive video systems typically draw audio and video (separately or together) from a medium such as a 30cm/12" LaserDisc, and computer-generated control programs, text and graphics from a microcomputer such as an IBM Personal Computer (PC) or Apple Macintosh. All these data are processed by the system, usually through the use of one or more interface boards (or cards) and related software.
- Multimedia systems integrate AV and computer resources on a single medium, which may be some form of 12cm/4.75" compact disc (CD) or computer disk (a convenient distinction of media by spelling). Processing power, too, is integrated within a single box, which might be a player dedicated to a specific CD format, or a microcomputer enhanced with special hardware and software resources.

In conceptual terms, this presents the user – who might be a student, a trainee, a shopper or a museum visitor – with an element of personal control over the flow of the programme. The user may ask or answer a question, indicate a choice or make a decision: the system responds by presenting information which meets each user's perceived needs or interests. A new generation of hypertext and hypermedia products takes this concept one stage further, by establishing automated links between related words or themes, so that users can freely pursue ideas anywhere within a given body of information.

PART 1: TECHNOLOGY

Interactivity is characterized by four basic concepts:

- Rapid random access, which ensures that information of all kinds can be retrieved both quickly and accurately.
- Branching, which allows the user to follow specific lines of inquiry away from the main trunk of the programme into areas of special interest.
- Transparency, which ensures that the workings of the technology do not distract or inhibit the person in front of the screen.
- Navigation, which provides the means for the user to move easily and confidently within the interactive programme.

Interactive systems have many advantages. They can provide a large body of information from many sources – sound and pictures, still and moving images, computer text and graphics – in a single disc no larger than an LP or CD. They can deliver the same basic information in the same way every time, and provide a wealth of additional material to meet the needs, interests and abilities of many users. Well-designed interactive systems are interesting, responsive, fun and easy to use. They can be relatively inexpensive to buy, and are usually more reliable and cheaper to maintain than videotape, tape/slide or other conventional AV systems.

In any context, interactive systems dynamically present objects, ideas and situations which are difficult to convey through static or linear media, and allow users to explore, study and compare information creatively and constructively. Museums and galleries are well placed to exploit this new technology, both as a means of presenting collections which are too large, diverse or fragile for conventional displays, and of publishing scholarly work, reference material and resources for professional, educational and consumer markets.

A typical delivery system also includes a monitor, and one or more input devices such as a keyboard, mouse, keypad, touchscreen, light pen, smart card or barcode reader; public information systems particularly may also include a printer of some kind. Consumer CD products are designed to connect to ordinary TV and hi-fi systems, or home computers, but IV and multimedia products for professional markets may require multi-purpose monitors, touchscreens or tamper-proof hardware and housing.

There is a wide choice of rapidly-developing media and platforms. In the early 1990s:

- LaserDisc provides excellent full-colour, full-screen moving pictures with sound – video as most people expect it – but interactive delivery systems can be expensive and cumbersome by comparison to some emerging technologies.
- CD-Digital Audio (CD-DA) is an excellent medium for high-quality or large-volume sound recordings, and the discs can be played in all other types of CD players.
- CD Read-Only Memory (CD-ROM) is a data storage medium capacious enough to hold all the published names, addresses and telephone numbers in the US on two discs; it has limited facilities for audio, graphics, animation and stills.

- CD-ROM eXtended Architecture (CD-ROM XA) has additional facilities for integrating audio, stills, graphics, animation and text displays, and is compatible with both CD-ROM under an IBM or compatible PC, and CD-I.
- CD-Interactive (CD-I) was released as a consumer product by Philips in the US in October 1991, and in Europe in April 1992. It carries audio, stills, text, graphics, animation and computer data, with its own dedicated player, and a catalogue of software titles.
- Commodore Dynamic Total Vision (CDTV) was launched as a consumer product by Commodore in the US and Europe in April 1991. It also supports audio, stills, text, graphics, animation and computer data, with a system based on the popular Amiga computer (some models of which can be upgraded directly). Both CD-I and CDTV were launched with limited capacity for moving video, but both promise to upgrade to full-screen, full-motion video (FMV).
- Digital Video Interactive (DVI®) technology is a process rather than a product, a set of chips for a desktop computer which compresses and decompresses still and moving pictures for use within a computer and storage on CD or disk.
- Various other forms of compression and decompression are emerging, and the recommendations of the Joint Photographic and Motion Picture Experts Groups (JPEG and MPEG) to the International Standards Organization (ISO) have already established *de facto* standards to which many manufacturers are producing hardware and software products.
- Leaders in the professional computing market now offer fully integrated multimedia computers and tools, including IBM (with the Ultimedia line, the Audio-Visual Connection, M-Motion Video and other products), Apple (with QuickTime, HyperCard and other products), and the Multimedia PC (MPC) group initiated by Microsoft (with a specification, upgrade kits and trademarked products).
- High Definition Television (HDTV) produces TV and video images of cinematic quality, by doubling current broadcast standards, and changing the aspect ratio of the screen from 4:3 to 16:9; however, two HDTV standards (one European, one Japanese) are already in competition, and both demand new hardware from broadcasters, producers and viewers.
- Photo CD will be launched by Kodak in 1992, for consumers and professional applications such as desktop publishing (DTP) and multimedia production. The disc holds up to 100 images from 35mm film, at quality up to four times that of HDTV, for display on an ordinary TV, or use within a computer-based system, and should soon also carry audio and video.
- Holograms offer a rather different use of laser technology in high tech displays, and an unusual commercial publishing opportunity.
- Dark rides and 'experiences' reconstruct historic sites, or create imaginary scenes, using a variety of media and effects to enhance sight with sound and even smell. These displays are typically found in theme parks and similar tourist attractions, but have also been integrated into museum galleries.

As well as technical specifications and media mix, any organization new to interactive systems and multimedia must also consider a number of factors critical to the successful implementation of any new technology. These include cultural and ergonomic issues (which affect the design of software, systems and housing), as well as practical questions of choosing suppliers, providing technical support, and planning upgrade paths for hardware and updating routines for software.

PART 2: APPLICATIONS

The companies and institutions discussed in Part 2 of this report offer some useful models and advice:

- The Bank of England Museum uses IV to illustrate rooms which cannot be opened to the public (including the Gold Vaults) and introduce concepts such as Banking Supervision, which would be difficult to convey engagingly in conventional displays.
- Birmingham Museum uses IV to consider objects in *Collectors in the South Pacific* from the perspectives of a missionary, a tourist, a Melanesian museum curator and the collector himself, a British-born businessman. The system also offers an image base which can be searched by criteria such as provenance and media, as well as themes such as 'faces' or 'the sea'.
- The British Golf Museum in St. Andrews uses CD-I and LaserDisc to support AV presentations including games and databases, and has adapted material from these displays for publication on CD-I as home entertainment.
- The British Museum Education Service, with the British Library and Oxford's Computers in Teaching Initiative (CTI), developed an information package on Anglo-Saxon history, art and culture for use in *The Making of England* temporary exhibition, and subsequent enhancement as a CD-ROM publication for schools.
- Cambridge Interactive adapted the Cambridge Database System 2000 to support image bases on LaserDisc with a computer database which has been used to create a multimedia resource on the Naga tribes of Assam.

- Cartwright Hall, Bradford, uses IV with computer text which can be updated periodically, and a separate text display which museum staff can update daily with news of special events.
- The Design Museum in London features a game that attracts one visitor in five to *Design a Toothbrush* on a PC-based IV package, and the *Study Collection Database*, which combines text and digitized pictures on an Apple Macintosh.
- The Imperial War Museum uses IV to complement displays with documentary footage, and to show vehicles and aircraft in action. The museum also has two 'experiences' which simulate scenes from the trenches and the Blitz.
- Jersey Museum used DVI technology to compile a database for visitors and residents, which will support over eight hours of sound, text, graphics and moving footage on a CD-ROM. The museum is also storing its entire photographic collection as digital images on computer disk.
- Manchester Museum of Science and Technology features an adventure game in a space capsule that introduces material from the National Curriculum to schoolchildren, and a public information programme on *Renewable Energy*, commissioned by the Department of Energy for its use at shows around the country.
- Musée Carnavalet used text and graphics from a computer with CD audio to introduce the music and memoirs of *Mozart à Paris*.
- The Museum of Civilization in Ottawa-Hull, the Museum of Mankind in London and a touring exhibition of the Canadian north all used versions of *The Living Arctic* to illustrate aspects of northern life with a set of interactive games and scenes of daily life.
- The Muséum d'Histoire naturelle in Paris used text, graphics and audio from a computer to provide information about birds, mammals and reptiles displayed as specimens behind glass.
- The National Museum of Natural Science, Taiwan, uses 'mini-theatres' to present videodisc-based shows on science topics, and will use IV to complement dioramas illustrating the world's ecosystems in galleries due to open in 1993.
- The National Gallery's Micro Gallery combines digitized full-colour images of over 2000 works with an illustrated art history database which visitors can explore freely to study and compare information, and request reference prints and customized maps of the galleries.
- The National Museum of Racing in Saratoga Springs, New York, is the subject of a case history which candidly describes some of the practical issues which virtually any multimedia installation must address.
- The National Waterways Museum in Gloucester uses IV in games which illustrate the challenges of building a canal, *Operating a Lock* and, with a real tiller and a life-sized video display, *Steering a Boat*.
- The Natural History Museum in London has several multimedia displays, including an interactive videowall, and two games which test memory skills. A temporary exhibition on the RSPCA used an illustrated trivia quiz to convey important messages about animal welfare.
- The Norwich Tourist Information Centre features a guide arranged as three thematic walks, which was researched and designed by local high school students, with technical support from the Norfolk Technical and Vocational Educational Initiative (TVEI).
- Royal Britain, a commercial exhibition in the City of London, provides a brief case study which emphasizes the importance of planning and coordinating interactive displays, and building good relationships with suppliers.
- The Royal Museum of Scotland produced two programmes in-house, with text and digitized still images. Both were extremely popular with visitors, and curators felt that the digitized images supported the interpretative material without upstaging the objects actually on display.
- Content for the interactive guide to *Glasgow's Glasgow* was prepared within a month by the exhibition's chief researcher, with support from suppliers who developed the interactive program on a Macintosh, and transferred it to PC-compatible Olivetti computers for the show.
- The Tate Gallery, Liverpool, has a dynamic database on modern British artists in *Sculpture Interactive*. The package includes facilities for users with visual and physical impairments, and for students and curators to create their own multimedia essays from resources on the LaserDisc and computer.
- The Victoria and Albert Museum in London features scenes of daily life and culture in *Images of China's Past and Present*, and offers unique views of objects in four media with magnified views up to 200 times actual size.
- The Wellcome Institute commissioned an image base on LaserDisc to complement a database on their mainframe computer, for use in their new museum by visitors, researchers and medical specialists. Other interactive exhibits will include information programmes and games.

Many other European and American projects are described briefly on pages 69 to 75. These include initiatives to catalogue and publish collections on disc, and to establish networks and consortia to share experience, expenses and resources.

SOME DEFINITIONS

interactive adj. 1. allowing or relating to continuous two-way transfer of information between a user and the central point of a communication system, such as a computer or television. 2. (of two or more persons, forces, etc.) acting upon or in close relation with each other; interacting.

Collins Concise Dictionary, 1988

interactive video – a medium that combines the power of the computer to control, manage and organize information, with the power of video to represent visual data.

Kerry A. Johnson and Lin J. Foa, *Instructional Design*, 1989

Interactive Video The convergence of video and computer technology: a video programme and a computer program running in tandem under the control of the person in front of the screen. In interactive video, the user's actions, choices and decisions genuinely affect the way in which the programme unfolds.

Signe Hoffos, *Interactive Video* (Parsloe, ed.), 1983

interactive video: Generic term for systems in which video, audio, and other elements are combined to instruct, entertain, sell, or otherwise influence a user who is permitted to determine a route through the system by interacting with it. The interaction may influence what is shown or what the system does, and may or may not result in some measurable change on the part of the user.

Peter Crowell, *Authoring Systems*, 1988

The convergence of computer and video technology to produce interactive video systems has meant that CBT (computer-based technology) is no longer restricted to text, and video need not be linear and passive .... A lesson based on interactive video may consist of moving pictures (*e.g.* archive documentary, drama), still pictures (*e.g.* slides, photographs, paintings, maps), computer graphics and text. This mixture of media motivates the learner and contributes toward the effectiveness of the training resource.

Ravi Singh, *Artificial Intelligence for Society* (Gill, ed.), 1986

A truly interactive instructional system is modular, individualized and responsive: materials are encoded on the disk in a modular, rather than a linear form; each learner creates a unique lesson sequence based on his or her own interest and learning style by selecting from hundreds of pre-planned options; as a learner engages the system in an interactive dialogue, the system responds with appropriate feedback, instructional sequences, or additional options that are suited to each learner's interest, ability, preferred style, pace and languages.

M. DeBloois, *Use and Effectiveness of Videodisc Training*, 1988

'Interactive' means that the viewer, not the programme producer, controls the sequencing of the programme. We are all used to linear programmes, which have a beginning, a middle and an end, and with which the viewer is merely a passive observer. An interactive programme on the other hand will pose questions to which the watcher can respond .... Depending on which response has been selected the programme will branch and show the next segment of the programme. This next segment could be anywhere on the disc or tape and not necessarily adjacent to the previous segment.

David Matthewson, *Revolutionary Technology*, 1983

Interactive AV [audio-visual] requires the active participation of the viewer. The participation itself can be on two levels. At the basic level, the viewer is simply making choices, using the random-access facility which is at the heart of any interactive system. More truly interactive systems require the viewer to give information about himself.

Robert Simpson, *Effective Audio-Visual*, 1987

In interactive video, the amount of interaction which can be exercised between the user and the IV system depends on a number of factors such as hardware (*e.g.* the type of video player), software (*i.e.* the instructions placed in the system in advance) and the overall design of the programme. But interactivity only takes place at all as a result of communications between the system and the user.

Gerard O'Neill, *Interactive Video in Training*, 1987

At this level, it is possible to merge the unique capabilities of the book, television and the computer into one integrated medium .... Applications are limited only by our imagination.

Efrem Sigel, *Video Discs*, 1981

We punish people by putting them in jail – we deny them interactivity.

D.P. Ely, *Interactive Media: The Human Issues*, 1989

BASIC CONCEPTS

One idea runs through all the definitions quoted on page 1: that interactivity is not a question of technology, but of human communication. Although much of the following discussion concerns hardware and systems, technology here is only a means to an end. It is important to understand how interactive systems work, but interactivity really has more to do with human interface than computer boards.

INTERACTIVE VIDEO AND MULTIMEDIA

The terms interactive video (IV) and multimedia represent a useful distinction, technically and historically.

IV came first, with many prototypes and demonstrations from the 1950s onward, and a range of commercial products from the late 1970s. IV combines some form of computing power (usually, a desktop computer) with some form of video tape or disc. By definition, true video – that is, full-screen, full-colour moving pictures of a quality at least equal to that of VHS videotape – makes up the larger part of the AV programme. Other components may include animation, audio and images of various kinds, and computer-generated text and graphics.

In practice, by far the most popular AV source is the 30cm/12" reflective optical videodisc known as LaserDisc or LaserVision. Properly, LaserVision defines the format originally developed and launched by Philips in the late 1970s, while LaserDisc is preferred for the current generation of discs (with better audio quality), supported in Europe by Philips, Pioneer and Sony and, in the US and Japan, by a number of other manufacturers as well.

The 1980s saw the emergence of the 12cm/4.75" compact disc (CD), and the enhancement of the computer's own facilities for storing and processing sound and images. Thus, the term multimedia was coined to describe computer-based systems which supply audio and some form of still and/or moving images from their own hard or floppy disks, from CD, or with a combination of media which may include videodisc or tape. Since true video demands a great deal of storage and processing power, multimedia systems often employ creative and technical stratagems to reduce the size, speed, quality or number of moving images in their presentations.

As technologies develop and converge, the distinction between computing, IV and multimedia is already blurring. It is less important, then, to understand particular technical specifications, than to appreciate basic concepts which describe all interactive multimedia systems – random access, transparency, branching and navigation.

RAPID RANDOM ACCESS

A video recording, like a film, is made up of many individual images, or frames. Any single one produces a motionless still image, while a series of minutely graduated images projected rapidly in sequence creates the illusion of movement. The human eye and brain need at least one-twelfth of a second to register a distinct image: thus, thousands of still images moving at up to twice that rate appear to us as moving pictures.

Every one of these frames can be numbered and indexed. These unique numbers can then be used to retrieve individual images or segments of moving footage, just as page numbers can be used to find individual references or chapters in a book. This is the rapid random access which makes so much else possible. Most LaserDisc players now offer random access to any image on the disc within three seconds, and some guarantee instant jump between images that are relatively close together.

A LaserDisc can hold some 55,000 frames on each side, every one of which can be identified by an unique reference number. In a simple programme such as the BBC's *British Garden Birds*, a printed index on the disc sleeve is sufficient to tell users that, of all the birds shown, information about the cuckoo starts on frame 27873. The user has only to punch those five numbers into the remote control keypad, and press the SEARCH key to retrieve that segment, in the very centre of the video recording.

Similar principles, of course, obtain for the wider world of computer-based information, so that any specific piece of datum can be retrieved quickly and accurately from the many records which the system may contain. The way in which computers actually handle information would appear frankly simple-minded in a human being, but this very simplicity allows computers to process large volumes of data very quickly and accurately. One key press from the user can bring together pictures from a videodisc, and text from a floppy disk, under the control of the computer's hard disk, virtually instantaneously.

TRANSPARENCY

In practice, the person in front of the screen is rarely aware of these myriad frame numbers, control codes

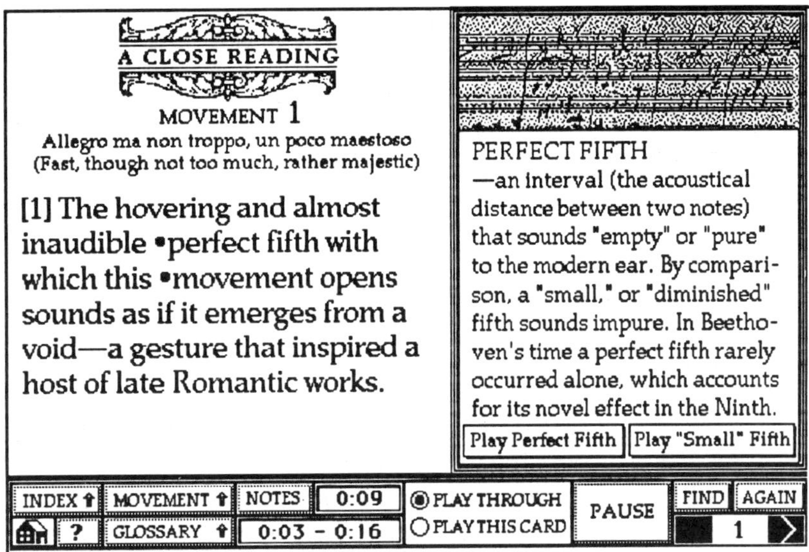

AN EXAMPLE OF INTERACTIVITY

The power and appeal of interactive information systems are illustrated in this screen from Voyager's *Ludwig van Beethoven: Symphony № 9.* The package combines an ordinary CD audio recording of Beethoven's Ninth with an illustrated hypertext package, on an Apple Macintosh or Multimedia PC (MPC) computer with CD-ROM drive.

The presentation on the computer screen combines text, drawings and animation to provide facts about the composer's life and times, an introduction to musical theory, an overview and a close reading of the symphony and a trivia quiz for up to four players. The package makes extensive use of audio from the CD and the computer itself. The close reading, with text screens like that shown here, provides running commentary as the symphony is played through headphones or speakers from the CD.

In the close reading, the main text appears in the left-hand window. Users can select any keyword (such as • perfect fifth or • movement) to call up a definition in the right-hand window: here, the explanation of a perfect fifth includes two short audio excerpts (provided by the computer) which demonstrate the difference between small and perfect fifths. (Another Macintosh package, *Practica Musica* from Ars Nova, provides ear training and musical theory entirely from the computer.)

At any point, users can select the options on the menu at the bottom of the screen to go to another part of the recording or text (and use facilities such as the Index and Glossary). All the information in the programme is accessible through links which make logical connections between related ideas and facts, so the user can pause within any section to check a definition or pursue an idea, knowing that the programme will return exactly to the point of departure, and resume the formal study in an instant.

and other data, for such detailed information is used only by the computing device in the heart of the system. The user simply presses a button or types a word to indicate a choice or request: the system itself interprets this as a command to retrieve specific information associated with that option.

This is the essence of transparency, that desirable end by which the user sees only what is of immediate interest and value, and all the information used by the computer, like the cables and wires which connect the various parts of the system, are kept well out of sight, physically and figuratively.

BRANCHING

The actual structure of an interactive programme is also much more complex than the person in front of the screen ever realizes – or, indeed, needs to know.

It is generally useful, even in a simple presentation, to give the user an overview of the information available there: it is as distressing to feel lost in an interactive programme as in any other unfamiliar environment. However, a table of contents or map of the programme is rarely more than a sketch of a complex infrastructure which supports a range of choices and possibilities.

Decision-making in such an environment is supported by branching – the facility which, as the word suggests, allows the user to move away from the main trunk of the programme to explore specific lines of inquiry.

Most training or information packages follow a direct route through main points which are built up one after another. In such programmes, branching is used principally to enhance the learning process by giving users the means to experiment with a variety of learning exercises.

A typical loop in a management training programme might present the trainee with a disciplinary problem, and suggest three possible ways of handling the situation. The trainee chooses one, sees the probable consequences of that approach, and gets a spot of advice before moving on to the next stage. However, the trainee can usually return to the decision point and explore the other options, learning something from each, before moving on. This combines an element of choice and the benefits of interaction with a lesson plan which ensures that every trainee sees all the basic material and at least a certain percentage of the optional information.

The same principle in a gallery display might take all visitors to a core of basic information from which they can branch out in search of details or interpretations which specifically interest them. Even in packages designed specifically to promote products to consumers, the same information may often be reached through a choice of routes, whether the shopper wants to go straight to a specific item, or be led there by a round of questions which identify the products most likely to be of interest.

The branch may be as simple as the fork between yes and no, or as complex as a simulation exercise which supports so many variables that no two users are likely to see the same sequence of events. In some exercises, every decision leads to another round of choice, in products that offer so many connections that the metaphor is less tree-and-branch than spider's web.

NAVIGATION

Like branching, navigation borrows a familiar notion from the physical world to describe a conceptual challenge in the fantasy land behind the screen. As interactive systems grow more complex, it is essential to provide their users with the information and tools they need to travel confidently within large bodies of diverse and unfamiliar ideas and images. The principles of navigation ensure that the user never gets tangled in the web, but moves lightly from point to point along logical strands of association.

Maps are, of course, essential. Initially, these were usually indexes or menus at the beginning of the programme, to which users could return at the end of each segment – an option which is still appropriate for many information programmes which address audiences with little experience of interactive systems. Where the volume of information or connections is greater, the main menu usually leads to any number of sub-menus – and a long and tedious journey from the most general to the most specific information, unless a few shortcuts are provided.

Thus, many interactive systems now offer the user instant access to a main menu which leads to all parts of the programme, and to resources such as dictionaries and indexes, from any part of the presentation. Keywords can be visibly distinguished from the rest of the text by colour or weight (that is, typographical devices such as UPPER CASE or *italics*) to provide instant access to background information, definitions, examples or other parts of the presentation. All these devices are illustrated in the screen from the *Beethoven CD Companion*, in the box on page 9: they include some conventions established by the archetypical HyperCard software package, such as the pictogram which leads to a main menu called Home.

Of course, it is sometimes advisable to limit access to some of these features during segments which demand the user's concentration – a training course, for example, may provide reference material but prohibit access to other parts of the programme while the student is engaged in a lesson, or conceal some resources during a test. These are some of the fundamental decisions which distinguish the creation of interactive programmes from all forms of linear design.

The Home icon so widely employed in HyperCard illustrates one important design trend – the use of visual and conceptual metaphors in place of conventional indexes and tables of contents. Thus, access to specific topics or tools may be effected through a simple pictogram – a book which represents the glossary, or a pencil which represents a drawing tool, for example.

Equally, the whole programme may be presented within an extended metaphor drawn from the subject matter itself. A gallery guide, for example, might present the main menu as a drawing of an entrance hall, where a Greek vase leads to information about the classical collection, and a Pop Art poster to the 20th century galleries. Visual metaphors alone can address an audience with a wide range of languages and reading skills; ideally, words and pictures together provide the verbal and conceptual keys to a wealth of diverse information.

THE INTERACTIVE ADVANTAGE

These points summarize what are widely perceived to be the particular, and sometimes unique, strengths of interactive multimedia systems (and, specifically, those which offer moving as well as still images).

- The same basic information is always presented in the same way, the hundredth time or the first.
- The system can hold a pool of basic information for all its users, and a variety of supplementary data for those who want or need it.
- Interactive systems can give their audiences the materials and the tools to explore, compare and discover information, features and structures for themselves, to ask and answer questions, study facts and relationships, and find advice and guidance.
- A well-designed package can meet the needs and interests of many people, and give them all the information which is best suited to their personal needs, abilities and interests.
- Multimedia systems present new facts and ideas in the ways in which we are most likely to remember them – by combining seeing, hearing, reading and doing, and by engaging the user in the active pursuit of information.
- Interactive media can employ a choice of instructional strategies and question types, to suit individual learning styles – the definable ways in which adults approach new information or tasks, and through which their interest and enthusiasm can most easily be engaged.

- Interactivity puts the user in control, and gives each individual a sense of ownership and responsibility for their own learning experience.
- The interactive system can use its processing facilities to record and analyse how its contents are used, to maintain records or collect data.
- Interactive systems are relatively novel, and potentially great fun – aspects of the technology which can be profitably exploited to engage the user's attention and provide a memorable experience.

Despite their many advantages, interactive media are not replacing people. If anything, they tend to provide information and training where none was offered before, and to complement other human and material resources. Experience has already shown that:

- Many people prefer to seek basic information and ask 'obvious' questions from a machine, which is perceived as both patient and non-judgemental.
- Thus prepared, users are more confident about approaching someone in a position of authority for additional information or services.
- Users can gain basic information from the system before engaging the time and attention of specialists: in many professional or commercial environments, this makes the best use of everyone's time and skills.
- By sharing the routine (and often frustrating) tasks of basic instruction with the interactive system, qualified staff are free to concentrate on what people do best – dealing with individuals on a personal level.
- When human resources are overstretched, the interactive system can provide those who are waiting for personal attention with practical information which is relevant to the topics they wish to discuss.
- The system is beyond the moods, prejudices and personal stress which can affect the performance of any guide, lecturer, trainer or salesperson.

Interactive media have the resources to create a rich and varied learning experience, in museums and shopping malls as well as schools, colleges or training centres.

Journals such as *The Multimedia & Videodisc Monitor* regularly present key findings from formal evaluative studies, and the books and journals cited on pages 80-81 include several which discuss the efficacy of interactive systems in learning. But time and again, formal reports and anecdotal observations point to this combination of senses and stimuli: people tend to learn more, and remember it longer, when they work directly with interactive information systems.

A BEGINNER'S GUIDE TO MULTIMEDIA

Commodore Dynamic Total Vision (CDTV) • A consumer product, for use with an ordinary TV or Amiga computer, developed by Commodore. CDTV offers graphics, text, stereo sound and moving footage.

Compact Disc-Digital Audio (CD-DA) • The CD format for top-quality audio, which holds around 72 minutes of sound on one side.

Compact Disc-Interactive (CD-I) • A consumer product, for use with an ordinary TV, developed primarily by Philips. CD-I offers audio at four levels of quality in stereo or mono, four formats for photos, graphics and animation, plus moving footage.

Compact Disc-Read Only Memory (CD-ROM) • The CD format developed for computer data storage, which holds at least 650MB (some 250,000 pages) of text and data, with limited audio and graphics.

CD-ROM eXtended Architecture (CD-ROM XA) • The enhancement of CD-ROM which improves the handling of multimedia packages.

Digital Video Interactive (DVI) • A technology developed by RCA, General Electric and Intel (with IBM, Matsushita, Olivetti and others), to process video in a computer, for storage on hard disks and optical discs. DVI technology currently supports three video formats for still and moving images.

Erasable Programmable Read-Only Memory (EPROM) • Data storage devices which can be erased and overwritten (for example, to run small programs in a system which employs a microprocessor rather than a separate computer).

Hypertext • Software programs which allow users to explore thematically-linked databases through the use of keywords, searches and/or visual metaphors such as icons and objects. **Hypermedia** extends this principle into multimedia environments.

Interactive video (IV) • Information systems which typically combine sound and pictures from an analogue video disc or tape with the text, graphics and processing power of a digital computer or microprocessor, in a delivery system which also includes a monitor and input devices such as a keyboard, mouse and/or touchscreen. **Multimedia** systems typically employ a single software source (usually a CD), a one-box delivery system, and all-digital rather than analogue/digital data storage and processing.

Megabyte (MB) • A common measure of storage capacity, just over one million (2^{10}) bytes (the smallest practical unit of digital data storage).

Photo CD • The compact disc format developed by Kodak to store up to 100 images from 35mm film, in a choice of quality levels, for display on an ordinary TV or use in a computer-based system.

Read-Only (RO) • Media on which data, once written (recorded) cannot be amended or erased.

Video Home System (VHS) • The most popular 12.5cm/0.5" videotape format.

Write Once, Read Many (WORM) • Media on which data can be recorded once, but not erased or overwritten subsequently.

AUDIO & VIDEO

Multimedia systems can combine a range of aural, visual and textual effects not practicable in any other single presentation. These may include:

- Live action (recorded, unscripted, as it happens), re-enactment (based on real events), dramatization (realistic but fictional) and simulation (realistic responses in hypothetical situations, possibly using real tools or equipment).
- Scripted words presented as text, captions or voice-overs, or by a presenter on the screen.
- People in all fields from professional presenters and actors delivering a script, to subject experts and specialists, and representatives of the audience itself.
- Visual information, presented as custom-made stills or video, computer graphics, or archival material such as maps, drawings, cartoons, paintings, posters, photos, slides, charts, graphs, technical specifications, blueprints and historical documents.
- Material transferred from photographic media such as 16mm or 35mm film, professional-quality videotape, 35mm slides and Ektachrome.

In an interactive presentation, a single videodisc or CD can support many hours of tailored, self-paced research or study, especially when the audio-visual programme is enhanced by complementary computer text and graphics.

STORAGE CAPACITIES

It may be natural to wonder how long a multimedia programme lasts, or how much a CD holds, but the answer is inevitably 'How long is a piece of elastic?'. The various storage media introduced on page 11, and described more fully in the following pages, each have finite storage capacities – typically, about 110,000 frames of video and two hours of audio on a LaserDisc, or 650MB of data on a CD – but the type of sound and pictures stored there, and the use of this material by the interactive control program, ultimately determine how much information any one disc may hold, and how long any one presentation of that material may last. A conventional audio CD holds at least 72 minutes of top-quality sound; the same disc differently recorded could store 250,000 pages of data, an encyclopædia with sound and pictures, or an adventure game that never ends.

A LaserDisc typically holds some 55,000 frames per side. This is enough for at least 30 minutes of high-quality, full-screen video – which is not much, by comparison to ordinary videotape. However, the judicious use of text screens, still images, computer graphics and animation can effectively extend the capacity of a single disc well beyond the hour or so that the AV material might take to play from start to finish. A range of financial training packages from Price Waterhouse and Premiere Productions, for example, typically support some 10 hours of study and exercises with one videodisc and two floppies, by complementing video sequences with computer text and graphics.

The LaserDisc has two audio channels, which can be recorded together for at least an hour's worth of stereo, or separately for two hours of mono. The careful deployment of both tracks can also extend the choice of information, and the ultimate storage capacity, of a single package. In those same financial training programmes, one track accompanies video from the disc, while the other provides extra voice-overs to accompany (and enhance) the many screens of computer text and graphics.

Multiple audio tracks also mean that a well-designed disc can present the same visual material to different audiences – for example, in different languages, at different levels of comprehension, or from different perspectives. Compact discs offer up to 16 separate tracks at a time – one training package (for TetraPak, on the CD-ROM XA format described below) offers 12 languages on one disc.

UPDATING AND AMENDING

Multimedia systems typically have the advantage of storing the entire interactive programme on one medium, where an IV programme is usually divided between LaserDisc and computer disk. The CD or LaserDisc, once recorded, is sealed tight; a computer disk may be protected against unauthorized access, but can usually be changed or amended at any time. It can be useful to prohibit or at least restrict users' access to the heart of a consumer product, but there are certain advantages in allowing for changes to professional presentations, particularly in fields such as exhibition design, marketing and training.

Independent text displays allow authorized staff to change volatile data such as prices and dates, and to add new or temporary information where appropriate. Systems linked in a network can even be updated centrally. In an IV system, storing computer programs and text separately allows a safety margin to update or amend the presentation without having to press a new disc. One tidy solution (adopted by Cartwright Hall, Bradford) complements the IV display with a simple electronic billboard: the main system is formally updated at six-monthly intervals, but local staff can add news and special events to the billboard at any time.

Updating the AV element is another question. With a programme stored wholly on computer disk, data of any kind can be added or deleted at any time, although the efficacy of this usually depends on both the flexibility of the original design and access to the appropriate production tools. Data once recorded on a standard LaserDisc or CD cannot be erased or re-recorded (although, at a pinch, a separate computer program might be written to prohibit access to unacceptable material).

There are recordable formats for both LaserDisc and CD, but these are better suited to the gradual accumulation of an image base than to the multifarious demands of an interactive multimedia presentation. Anyone commissioning or designing an interactive programme must seriously consider the need for updating and amending, to ensure that the programme is designed and executed to enjoy the longest possible shelf life.

INTEGRATING ANALOGUE SIGNALS AND DIGITAL DATA

One critical difference between IV and multimedia in the early 1990s lies in the use of video, and the way in which the various components of the presentation are stored and processed.

IV typically combines AV material from a videodisc with text and captions from a computer – frequently on the same screen, so that the computer text provides captions for the video, or the disc supplies windows of moving footage within a computer-generated display. This involves two fundamentally incompatible types of signal: the continuous, constantly varying signal of the analogue AV recording, and the discrete computer codes of the digital computer data

At the heart of any such system lie one or more interface cards (or boards), which may convert all the signals to one kind or another, or synchronize the two for display on a monitor which accepts both. In the early 1980s, synchronizing video and computer signals was the biggest hurdle on the road to interactive video; now, many commercial products, hardware and systems suppliers, and experienced programme-makers provide the requisite expertise and equipment. It is necessary for those who commission this work simply to understand the issues which inform fundamental decisions about the choice of media and delivery system.

The IV equation is further complicated by the fact that analogue video is tied to the technical standards

PAL, NTSC AND SECAM

For historical and political reasons, different and incompatible methods of composing a television signal have been developed and adopted in different parts of the world. These systems are generally known by names which actually describe only the composition of the colour signal:

- PAL (Phase Alternate Line).
- NTSC (National Television Standards Committee, also known as Never Twice the Same Colour by its detractors).
- SECAM (Sêquential couleur à memoire).

Each combines three standards to create a television or video picture. The line standard determines the number of scan lines used to build up a single screen. NTSC now has 525 lines per screen, while PAL/SECAM have 625 (and, thus, produce a finer picture). Anything from 405 to 819 lines have been used in various TV systems.

Each frame of 525 or 625 lines comprises two interlaced fields. The even-numbered lines are scanned in one pass, the odd-numbered lines in the next, and the two field scans are interlaced on the screen. These single frames are then run at speed, so quickly that we perceive motion in what is really only a rapid sequence of still images.

The field or frame standard usually relates to the frequency of the national mains power supply. The PAL/SECAM field standard (developed where the power frequency is 50Hz) is 50 fields, or 25 frames, per second. In NTSC (developed where the power frequency is 60Hz), video runs at 60 fields, or 30 frames, per second. (Standard film now runs at 24 frames per second – and special processes exist to add or subtract the requisite number of fields and frames to reconcile material from film to TV, or between PAL/SECAM and NTSC.)

The colour standard has to do with the way in which the basic colour signals are recorded and transmitted. This depends on chrominance (hue and saturation) and luminance (brightness), signals which are composed and broadcast separately. (The human eye also processes images and colour separately, and we are more sensitive to changes in brightness than colour – a fact which informs some critical decisions in screen design.)

The main difference between PAL and SECAM is the handling of the colour signal, so some PAL and SECAM video material is interchangeable if viewed in black-and-white. PAL is basically a refinement of NTSC, but uses more complex receivers to produce a more stable colour picture.

Multi-standard hardware offers immediate access to all three formats. Radical innovations such as digital video and high-definition video (HDV) offer much better images and obviate incompatibility altogether. These products are available in the professional market now: the challenge is to convert some of the world's billion-odd TV sets (not to mention video equipment and broadcast television facilities) to any new standard.

of the national broadcast television service (described on page 13). The salient point is that Britain and most of Europe employ the PAL standard for television and video, while France and much of Eastern Europe and the erstwhile Soviet Union use SECAM, and the US, Canada and Japan support NTSC. PAL and SECAM employ the same basic principles, moving at a full speed of 25 frames per second, but handle colour differently; NTSC, at 30 frames per second and with yet another colour process, is wholly incompatible with either PAL or SECAM. Thus, an NTSC LaserDisc or videotape will not run on a PAL player, and *vice versa*.

Hardware provides an immediate solution in the IV and video markets, with multi-standard videotape and LaserDisc players. Multimedia platforms take a radically different approach, by converting all source material to digital data during the production process, and delivering both AV and computer software in one package (typically, on CD or disk), for use with ordinary television receivers, video monitors or computer screens.

Consumer products such as CD-I and CDTV are designed for use with the family TV, home computer and even hi-fi – the delivery system itself converts the digital data on the disc for output to screen and speakers. Similarly, products for professional and educational computing markets, such as CD-ROM and DVI technology, can employ any computer with the appropriate software, storage and processing power, and a monitor capable of displaying their high-quality graphics.

This produces multimedia packages which are as widely compatible as any other form of computer software. No single medium is yet universal – a CD-I disc will not run in a CDTV player any more than a PC disk will work on a Macintosh computer – but any CD-I disc will run in any CD-I player, no matter where either was made. Only the power cable to the mains (which, again, is tied to a national grid) must be adapted to local conditions.

DIGITAL VIDEO, COMPRESSION & DECOMPRESSION

Multimedia systems clearly have advantages; however, video – at 25 frames in PAL/SECAM, and 30 in NTSC – demands a vast amount of data to be retrieved, processed and output every second. Media such as LaserDisc and videotape were conceived expressly to do this; computer media such as hard disk and CD were designed primarily for text and data, expanding their capacity for audio, graphics and, ultimately, animation and video only by stages.

The CD that can hold 250,000 pages of data could not store a minute's worth of video, using conventional techniques. Thus, where any volume of true video is required, a LaserDisc is still a sensible and cost-effective choice. However, much effort has been expended to reduce the volume of data demanded by video. Compression systems eliminate redundancy in the video signal, and store the least amount of information necessary to decompress the image when it is reconstituted on the screen.

In analogue or digital environments, an image is composed of many discrete lines; digital systems divide each line into a uniform number of discrete picture elements (pixels). The more lines per screen, and the more pixels per line, the finer the detail; the more colours available, the more subtle and natural the shading. An ordinary PC with industry-standard VGA graphics has a resolution of 640x480 and a choice of 256 colours from a palette of 262,144, but many graphic formats support the 16,777,216 hues available on more powerful display systems, and Photo CD offers resolutions up to 2048x3072.

Compression systems analyse patterns in this mosaic. At a rate of 25 or 30 images per second, the difference between one frame and the next is often slight: *interframe* compression systems typically record all the data in the first frame in a sequence, and then note only the changes from one frame to the next. Equally, the data in a single frame can be analysed by *intraframe* compression systems which effectively describe, rather than literally record, the details of each image – not noting every pixel in every line, but only the position and colour of the first, and the points at which the colours change thereafter. This is enough to describe the picture to a complementary decompression system, which reconstructs a coherent image on the screen.

Compression systems are discussed again on pages 23-24. Most, in the early 1990s, trade capacity for quality in the storage and processing of digital images. To provide video at all, the first generation of multimedia CD systems have been restricted to partial-screen displays and partial-motion speeds of 12 or 15 frames per second – windows of movement within the screen, which are often perceptibly grainy or jerky. But compression systems are improving rapidly – in March 1992, Voyager released the first CD-ROM discs to store feature films (using Apple QuickTime) for $29.95 apiece. CD may soon rival videotape in the consumer market.

Equally, multimedia developers are learning to play to the strengths of the technology – not the least, to abandon the model of broadcast television (which makes prodigal use of moving footage to answer its own technical limitations) and design interesting programmes with radically different values.

In the meantime, it is interesting to observe how many techniques characteristic of computer-based systems – windows, icons and text overlays – are eagerly imitated in other media, particularly television. Multimedia need not look like television: if anything, TV now wants to look like multimedia.

HARDWARE & SOFTWARE

All computer-based systems, from complex flight simulators to simple information terminals, depend on a combination of hardware and software. Machines and equipment are hardware, the computer programs which control them, and the AV material which they present, are software.

In multimedia systems, the combination of equipment or hardware which actually presents the interactive programme is often called a delivery system. This typically comprises:

- A source of processing power, which is usually a desktop or personal computer (PC), but could also be some form of microprocessor.
- Interface cards (or boards), built into the computer, which enable the various parts of the system to communicate with one another.
- A large-volume hard disk, which is usually built into the computer, often with one or two drives into which floppy disks can be loaded.
- One or more sources of audio-visual material, such as a videotape, videodisc and/or CD.
- A monitor, which usually displays still and moving video pictures and computer-generated text and graphics, and delivers audio through mono or stereo speakers.
- Other output devices, such as a printer or headphones, where these facilities are required.
- An input device of some kind – typically, a keyboard, keypad or touch-sensitive screen – which allows both programme-makers and end-users to interact with the system.

Multimedia software typically combines still and moving pictures with computer text, graphics and control programs. Two useful distinctions in spelling emerge here, between the computer *programs* stored on hard or floppy *disks*, and the audio-visual *programmes* stored on optical *discs*. The former provide the instructions to the system which make it work, the latter comprise the multimedia presentations which the end-user actually sees.

The distinction between hard and soft is confusing, for computer disks and videodiscs are tangible, but their value lies in their intangible contents, software. Thus, blank storage media are often described as consumables. In training, it is common to describe the whole package – videodiscs, computer disks, workbooks and manuals – as courseware.

It was also easier to distinguish the computer program from the AV programme when the former could produce little more than crude text and line drawings, and the latter demanded the production facilities of a professional video unit. Since then, the transition from IV to multimedia has been effected largely by the dramatic improvement in computer graphics and digital video facilities. It is now possible to capture and manipulate still and moving photographic images entirely within a computer, and to store a whole multimedia presentation on a CD – text, graphics, audio, video and control programs. The practical aspects of these developments are discussed more fully in the pages which follow; the significant trend is toward ever more compact, integrated and versatile media and systems.

MULTIMEDIA AND IV DELIVERY SYSTEMS

A CD-I or CDTV player, or a wholly computer-based display, typically comprises one neat box (about the size of an audio CD player), plus a monitor and some form of input device. At home, the monitor is likely to be the TV or home computer, and the input device will probably be a hand-held remote control unit or the computer's keyboard. In public, the monitor is likely to be a rugged and tamper-proof professional unit, and the input device may be a keypad built into its housing, or a touch-sensitive screen. The software within the system is usually a compact disc or, in a wholly computerized environment, a hard or floppy disk.

IV systems tend to be larger, with both a LaserDisc player and a computer, or perhaps a single box with drives of both kinds, plus an appropriate monitor and input device. In training and education, the keyboard is widely used; in exhibitions and information services, a touchscreen or keypad tend to be simpler to use and less vulnerable to damage. With IV, the computer software is often stored on the computer's hard disk – again, both for convenience and to save fragile floppy disks from wear and tear. The LaserDisc is resilient enough to stand up to frequent handling in the classroom, but can simply stay in the player until there is occasion to change one disc for another – which, in the case of permanent displays, might not arise for months or years, until the content of the programme is changed.

COMPUTER PLATFORMS

Multimedia systems can employ virtually any form of computing power, including the microprocessors built into some LaserDisc players. In practice, the market for IV in training is dominated by the

IBM PC and clones from other manufacturers, but displays in museums and galleries are as likely to use an Apple Macintosh, or even CD-I – all separate, incompatible computer platforms. Furthermore, the success of the special products for schools (such as, in the UK, BBC Micro, RM Nimbus and Acorn Archimedes) has created a niche market there, and the Commodore Amiga has another niche in video production facilities.

The catch is that models in the same range often have significantly different specifications, and products from rival manufacturers may be based on wholly incompatible principles. Thus, both hardware and software must be developed, rather like car parts, for different makes and models. This sometimes restricts the choice of ready-made products, and profoundly influences the development of customized delivery systems.

The contrast between the two main trends in computer design might be characterized as the difference between programmers and end-users. The traditional approach – epitomized by the Disk Operating System (DOS) characteristic of IBM and compatible PCs – makes extensive use of cabalistic codes, words and concepts unique to computing, which the user must input with letter-perfect accuracy at the computer keyboard. This is a logical extension of computer programming, but demands considerable training and experience of the end-user.

A different approach, characterized initially by the Apple Macintosh and now by PC products such as Microsoft Windows, makes extensive use of verbal and visual prompts, words and concepts drawn from everyday life. Users typically select information and options from multiple-choice menus, or employ visual metaphors such as icons, which automate many activities so that one action from the user initiates a chain of activity within the computer.

The computer screen here becomes a metaphorical workplace, which users can arrange even to the point of sorting files in folders and discarding unwanted documents into the 'trash' or 'wastebasket'. Skilled operators still require both training and experience, but new and casual users often find such systems easier to use than those which depend largely on unique codes and keywords.

This fundamental change in perspective, from the demands of the computer to the needs of the user, is reflected in the range of tools which now allow people without formal programming skills to create their own interactive presentations. These include both software products to design text and graphics, and create computer programs, and hardware products to capture sound and pictures from other media for use in computer-based systems.

Another useful distinction reserves the term PC for the IBM range and its clones, and uses the term desktop generically to describe any microcomputer. Old IV hands often describe products elliptically, so that a '3000 card', for example, is VideoLogic's Interactive Video Adapter (IVA) 3000, a 'DVA 4000' is their Digital Video Adapter, and a 'MIC system' uses complementary Multimedia Interactive Control software. Similarly, the numbers 286 and 386 refer to industry-standard processors in IBM and compatible micros.

CHOOSING A PLATFORM

The upshot of these developments for new users is that there is now a real choice of platforms and products to suit a market which is growing in size and diversity. The ultimate choice involves many features, including the various input and output devices described in the next section, the quantity and quality of AV elements required, and the scale and scope of the installation as a whole.

The 1990s have seen a new spirit of cooperation between IBM and Apple (who are now developing products together), and much raw data can be exchanged between fundamentally different systems. This is a great help in professional environments, and some multimedia developers employ PCs, Macs and Amigas variously during the production process. However, for the delivery of the finished presentation – and the economical development of programming in-house – it is essential to choose wisely from the variety of products now available.

A number of the displays described here were developed and delivered entirely on a Macintosh. Most IV systems employ a PC, but the Mac and other platforms are gaining popularity here, too. New products for the consumer market, such as CD-I and CDTV, offer versatile and inexpensive systems for use in museums and galleries. Moreover, 1991 saw some significant new products: Apple QuickTime provides developers with tools to adapt virtually any Macintosh software to multimedia; the Multimedia PC (MPC) initiative, led by Microsoft, provides a specification, upgrade kits and trademarked computers to bring multimedia to PC users; and IBM's own Ultimedia line enhances its PS/2 with CD-ROM XA and other features for multimedia. Where IV was often perceived as a variation on video, multimedia is rapidly becoming an integral aspect of computing.

No one system is ideally suited to any and all environments and applications, for interactive multimedia displays range from illustrated databases to the virtual reality of 3D simulators. Factors as diverse as designers' previous experience and preferences, and the prospects of corporate sponsorship may influence any one choice of hardware and software. Other considerations are raised in turn below, and discussed collectively on pages 36-37.

INPUT & OUTPUT DEVICES

In training, the most common input devices for IV and multimedia systems are the keyboard and mouse, both of which are now familiar in professional computing environments.

The mouse is less familiar, and initially a bit perplexing, to people who do not regularly work with computers. A mouse is a small hand-held device, connected to the computer by a long thin tail, with a freely rotating ball mounted on its underside, and one or more buttons on top. Moving the ball about on a flat surface such as a desktop (ideally, on a smooth spongy mat called a mousepad), and confirming choices or decisions by clicking on the button/s, directs an icon or cursor on the screen to select information, indicate choices or initiate activity.

A number of other input devices have been developed for professional and consumer markets, and the particular demands of public information systems. These include:

- A touch-sensitive screen (touchscreen), which allows the user to point to text or pictures on the screen to make a choice or decision.
- A keypad or remote control handset, similar to those for the TV or VCR, but usually custom-made to offer options relevant to the programme on display.
- A trackerball (or rollerball), which works like a mouse but has both the rolling ball and the button/s mounted on top of a small fixed unit.
- A joystick, popular in computer games to control the movement of an object on the screen with a short stick which rotates in the socket of a fixed mounting, with one or more buttons to effect choices and actions.
- A light pen, which selects information by touching a word or image on the screen, and may even allow the user to write or draw there.
- A barcode reader, like those in shops, which uses a light pen or fixed reading device to interpret a pattern of thick and thin lines and spaces (typically, printed on a label, control panel or workbook) as coded instructions or data.
- A concept keyboard, which is actually a touch-sensitive panel which can be customized with a variety of interchangeable overlays.
- A smart card, like those used in cash dispensers, which can hold data including security clearances for authorized users and simple control programs for computer devices.
- A customized control panel, superficially identical to that of a real piece of working equipment, which simulates the action of that larger, more expensive or vulnerable hardware.

Many systems use a combination of devices for different jobs: the keyboard to collect details such as the user's name, and the touchscreen to facilitate simple choices and decisions during the course of the programme. Some training programmes offer the user a choice of mouse or keyboard – it's simply a matter of personal preference, for the object is to make the user feel as comfortable as possible.

KEYBOARD AND TEXT SEARCHES

Two important factors governing the choice of an input device are computer-literacy and language skills. A computer keyboard offers dedicated control keys as well as a full set of alpha-numeric characters – as complete a range of options as any designer could want.

Where users are expected to be computer-literate (for example, in programmes for training or education) the keyboard is a convenient and familiar tool. Equally, where readers or students are likely to spend some time with the system, and need to develop appropriate skills – to manipulate a complex database, for example – the keyboard rewards those who learn to use it, by providing an ideal tool for free text search and note-taking.

However, keyboards depend on verbal facility and manual skills, which demand that users be able to read and type in the prevailing national language. Multi-lingual keyboards are complex professional tools, and pictograms of this subtlety are notoriously difficult to design. Equally, even where computer literacy is high, the very sight of a keyboard intimidates many. It is observable at computer fairs that even qualified operators are often reluctant to use unfamiliar keyboards – especially where they might be observed by strangers.

TOUCHSCREEN

Touchscreens are popular in public information systems. Anything which can be displayed on the screen can be employed to control the programme – users simply touch a word or image on the screen to make choices or initiate action. This provides ideal conditions for technical transparency and versatile navigation. The use of words and images on the screen easily establishes simple, consistent interface tools which gracefully adapt to different parts of the presentation, or different versions of the same programme.

One drawback is that many people are initially reluctant to touch a screen, for the experience is not only unfamiliar, but contrary to the common practice of training children not to touch the TV. However, touchscreens are becoming more popular as they become more familiar (and more reliable). Their principles are easily grasped, although thorough testing is advised to ensure that the system responds to a single, light touch within a reasonably large target area, and that every touch point leads to the appropriate destination. Users are quickly frustrated if the touchscreen demands a subtle technique, or only responds to repeated prodding.

Reactions are mixed to the relative virtues of capacitance and infra-red touchscreens. Both use x and y (*i.e.* horizontal and vertical) coordinates to identify the point at which the screen is touched. Capacitance systems require a firm touch to the membrane over the video screen to register a choice; infra-red screens interpret any break in the fine net of light rays over the screen as a touch. Sensitivity is valued particularly where screens are in constant use by qualified operators (for example, in engineering and other technical environments). In training and education, or displays for public use, it is often advantageous to find a screen which responds to a firm touch, but resists environmental interference.

Tony Blake of Intersearch Systems, who developed the proprietary interactive information system *Gardeners' Questions Answered*, reports that sunlight streaming through the glass roofs of the garden centres was strong enough to interfere with the infra-red screens; he switched to trackerball. Writing in *Audiovisual Librarian* [16(2)], Emma Davidson of the Imperial War Museum explained its choice of capacitance screens, noting that, with infra-red screens, 'it is possible in places on the screen to set the programme off without actually touching it, which may be off-putting'. (However, compare the Museum of Civilization's experience, on page 52.)

In the mid-1980s, Videodem studied the use of touchscreens in a point-of-sale system at Yorkshire Electricity Board showrooms, and found that older consumers particularly were wary of 'getting shocks'. At the same time, the EPCOT Center in Florida told *The Videodisc Monitor* (January 1984) that it had to provide more specific instructions than simply TOUCH THE SCREEN after observing that users were enthusiastically pressing their whole hands – and even their elbows – on the screen, and confounding the system by rubbing rather than pointing. The Carter Presidential Museum and Library in Atlanta, Georgia, reported the unique case of an over-anxious visitor's actually destroying a touchscreen; more typically, two dozen touchscreens provided sterling service, standing out of doors all summer during the 1988 Glasgow Garden Festival.

KEYPAD

The remote control keypad is now a familiar accessory to home entertainment products such as the TV, hi-fi or VCR. A keypad typically combines up to 10 numeric keys with an assortment of controls to effect at least the most popular functions on the player.

Good design here combines simplicity and functionality. A brief survey of keypad design in the videodisc industry amply illustrates the challenge of reducing complex operations to a manageable number of self-explanatory commands. It is a useful exercise for any designer of interactive products or programming simply to survey the range of keypads in a large electronics shop, to appreciate what some hardware suppliers expect of ordinary consumers.

Customized keypads typically employ 10 or fewer numeric keys and a small number of clearly labelled or colour-coded control keys. With careful planning, a few keys can control complex and interesting programmes – the question is creative rather than technical. Well-designed keypads are simple, reliable and user-friendly; with some foresight (such as the use of colour-coding or symbols rather than text), they can also support multi-lingual presentations, or be adapted to new programmes without redesigning the complementary hardware.

Research for the information system on *Family Credit* (made for the UK Department of Social Services by Vivid Interactive) revealed that typical users would be comfortable with something that looked like a hole-in-the-wall cash dispenser, but wanted as few keys as possible. That system was reduced to four large buttons, like those on arcade games (YES, NO and two directional arrows), and the housing redesigned to include graphics around the screen which invited passers-by to use the system.

MOUSE, TRACKERBALL AND JOYSTICK

A mouse, trackerball or joystick typically controls a flashing cursor on the screen, to select information and initiate action. Young children, or visitors unfamiliar with the principles of the device, often appreciate a large cursor which can be placed anywhere on a large target; where users are more skilled, the fine tip of a pointed cursor can neatly select one word from a long list, or one item from a busy screen. Again, the question is one of design rather than technology.

The trackerball and joystick, being fixed in their moorings, require less operating space than the mouse, and can be built into the delivery system's housing with their vulnerable wiring safely concealed. However, all three are vulnerable to tampering: chewing gum is a particular hazard in public information systems.

The principles of these devices are harder to grasp than those of touchscreens or keys, for a certain order of psycho-motor skills is required to relate the movements of the rolling ball or stick in the hand to those of an object on the screen. Rolling devices may be intimidating to some audiences; however, the popularity of the mouse now means that programmes for computer-literate audiences usually benefit from users' familiarity with the concept and the device itself.

Rolling devices are generally easier to use and less vulnerable to damage than keyboards, and better suited to fine discriminations than keypads or touchscreens. Where users are likely to spend some time at the terminal, or to return there frequently, it may be worth their while to master the unfamiliar device; training packages typically include an optional exercise for users who need to grasp the use of the mouse before starting the course.

A trackerball supported a versatile and responsive information system in *The Reuter File*, made for Reuters by EPIC Interactive Media to provide information about the international news service to staff, clients, shareholders and the public. Databases and retrieval systems of graduated complexity addressed the different interests and abilities of these various audiences. A simple visual metaphor, of folders in a filing cabinet, led directly to general information. A long, rapidly-scrolling text display presented an alphabetical list of topics designed for researchers accustomed to working with databases. Dense screens like those on dealers' terminals provided information to professionals from financial markets. A trackerball and buttons mounted on the purpose-built housing controlled the cursor, and effected choices and decisions.

LIGHTPEN AND BARCODE READER

Lightpens combine the precision of rolling devices with a tool which most users immediately understand. Pens and screens are familiar individually, although rarely used together outside the world of computer graphics. However, most visitors quickly grasp the idea of using a pen or stylus to point to words or pictures on the screen, or even to draw there.

Unfortunately, anything attached to a machine by a cable is vulnerable to damage, and the tip of the stylus can also be fragile. Lightpens are popular in specialized fields such as machine instrumentation – where professional operators learn to use them skilfully and carefully – but they are rare in public information systems.

Barcode readers can also employ a wand attached to the system by a cable, or a fixed reader. Most information systems seal the barcodes into a laminated booklet or control panel, as the wear and tear is more than ordinary paper can stand.

Barcode systems have gained some popularity in education and training, since they can support simple interactive programmes with no more than a microprocessor, and desktop publishing packages allow programme-makers to produce the barcodes themselves in-house. There is a small market here for reasonably-priced databases and resource discs which can be cheaply adapted for education by the production of barcoded workbooks.

Visitors to *Information Age* at the National Museum of American History in the Smithsonian Institution in Washington DC received a barcoded brochure which they could pass through scanners at various displays to activate interactive systems, record data about themselves (a useful market research tool) and get a souvenir print-out indicating the displays at which they had used their brochures. One version of the Tate Gallery Liverpool's *Sculpture Interactive* uses a barcoded workbook to control the multimedia database.

CONCEPT KEYBOARD

The concept keyboard is a popular alternative in education. It employs a touch-sensitive panel (usually A4 or A3 size) which is covered by an ordinary sheet of paper or thin plastic membrane, onto which words, symbols or any other prompts can be printed or drawn. The user simply touches the appropriate spot on the panel to select information or send instructions to the system. The panels offer many options to designers, and overlays can easily be changed to control different programmes from the same delivery system.

The features which make the concept keyboard popular in education apply equally to exhibition design. Large, colourful icons or pictograms are more inviting to touch, and easier to use, than devices which require skilful manipulation, advanced conceptual skills, or simply confidence in the face of new technology. Equally, where users need to spell out their requests or input data, a concept keyboard fitted with an alphanumeric overlay may prove simpler and sturdier than a conventional keyboard.

A concept keyboard was used in a system developed for Glaxo in 1989 by Interactive Media Systems (IMS) with The Soft Option and Interactive Meetings. The package was designed as a presentation aid for lecturers who were medical specialists rather than technologists. The touch-sensitive control panel combined basic operating functions to control the AV display, with a schematic map of the programme. Speakers simply touched the panel to call up illustrative material from the videodisc.

SMART CARD

Smart cards – which can record limited amounts of volatile information on a device which looks like a credit card – offer a practical combination of simplicity, security and record-keeping facilities. By recording pertinent data about individual card-holders, smart cards can control access to hardware or software, or monitor the use of systems or services for which customers can be charged.

The most familiar cards are used to bank at an automated teller machine (ATM), the hole-in-the-wall cash dispenser. The portable Centaur delivery system, which was designed for travelling sales reps, records simple IV programmes with text and graphics on individual smart cards. Some training systems use smart cards to control access to the large body of material in the delivery system, and to record each trainee's individual progress.

The *Life is Choice* section of the Denver Natural History Museum issues each visitor with a magnetic Life Card which, like the Smithsonian's brochure, activates various interactive displays and allows visitors to keep a personal health record. In the UK, the National Gallery uses smart cards to administer the sale of print-outs from the Micro Gallery image base. Visitors buy the card at an information desk (currently, at a rate of £1 per print-out), and select the images which they want from the interactive terminals. A card reader at the terminal deducts the cost of each print-out from the credit limit on the card.

OUTPUT DEVICES

As well as the most obvious output device, the monitor, some systems also incorporate a printer of some kind – a simple dot-matrix system for small notes or receipts, or laser, inkjet or similar high-quality device for larger, crisper print-outs or colour reproductions.

It is possible to grab a single frame from a video source such as an optical disc, and produce a good colour print at the delivery system. Some printers provide both A4 and postcard size images. This can represent a useful and even lucrative service for picture researchers, readers and visitors who want reference copies or souvenirs of images from special exhibitions or multimedia databases.

It is important to resolve the issue of copyright before offering prints or reproductions for sale or distribution (remembering that, even where the original object is in the public domain, the reproduction captured in the image base may represent the work of an independent photographer or picture agency). Unfortunately, controlling the use of images *taken* from such a system – which may then be reproduced elsewhere without permission – is still beyond the capacity of any interactive system.

An information system produced by VideoMedia for Coats Viyella, to promote Dorma bedlinen and decorating fabrics, offered a price list for each range on A4 paper. The system also included a facsimile machine, accessible only to authorized staff, which could be used to communicate directly to Dorma outlets in large department stores. The system was not working well on this writer's one visit to a Debenham's store in central London – neither a message on the screen, nor the silent laser printer indicated that a request for a print-out was being answered. Manual investigation eventually released a fistful of sheets stuck just within the machine itself; previous users presumably just walked away, thinking that the system had failed to deliver.

This is only too common a reaction to technology of all kinds, for it takes much less to disappoint the public than to please them, and the designers of anything so new and unfamiliar as interactive multimedia systems must strive to create experiences which – however simple – are engaging, rewarding and reliable.

CHOOSING INPUT AND OUTPUT DEVICES

Choosing input devices and designing a retrieval routine often represent a balance of interests between the demands of the interactive presentation and the abilities of the end-user. Where the programme is rich and complex, and the audience large and various, design rather than technology holds the key to a display which is interesting and informative, responsive and efficient.

One institution may use different delivery systems to present fundamentally different programmes, with different input devices – touchscreens or trackerballs in the entrance halls and galleries, for example, with keyboards and smart cards for databases and information services. Participative systems may even employ live video cameras, audio recorders and/or computer terminals to allow visitors to make their own contributions.

At the same time, many small establishments particularly favour the use of a single delivery system, which employs the same operating routines and visual conventions consistently no matter where or how it is used. There are, of course, advantages to both approaches: the ultimate choice should be made after a thorough and candid analysis of both the aims and objectives of the project, and the visitor profiles of the institution. Field trips to trade fairs, and to installations which already employ interactive displays, provide opportunities for observation and hands-on experience which can be invaluable in shaping an informed decision.

HOUSING

In any public place, most of the delivery system is held within a protective housing which displays only that which the user needs to view and interact with the programme. There may be an invitation or instructions on the housing or adjacent signage, an attractor loop on the screen, or simply a message such as TOUCH HERE TO START.

This message usually appears on the screen or button which activates the system for, when it has not, there have been any number of eyewitness accounts of users' vainly pressing inert labels on the housing, even clambering behind the system to reach the message on a display panel there.

Not surprising, then, that housing is a specialized field which excites as much discussion as any aspect of the specification. All housings must be secure and tamper-proof, and allow access to equipment for authorized staff, but their shape, size and decoration may vary from a wall panel to a free-standing kiosk.

Even where there are many temporary displays, it is usually most economical, and convenient for both designers and regular visitors, to provide permanent housings which reflect the same visual style and operating conventions. These must be designed to suit a variety of programmes and sites. A uniform system can be cost-effective in the long run, but planning must allow both time and money for field testing and modification.

Many design questions can only be decided by fitness for purpose. Should the delivery systems be free-standing, or built into other display units? At what height should screens and control devices be placed in an exhibition which may attract children and adults, people in wheelchairs, and visitors with a variety of skills, abilities and interests? Should there be a portable system, for use in meetings, seminars, travelling exhibitions or conferences?

Even with the best planning, observation will reveal unexpected responses from any system's real end-users. One designer reported, for example, that the very popularity of the children's material on one system discouraged adults from exploring other parts of the programme; eventually, the installation was adjusted to draw children to some terminals, and provide others for adult visitors.

CONTROLLING AMBIENT NOISE

Ambient noise in any public place easily reaches unacceptable levels even without audio information systems. Much can be conveyed without sound, but commentary and captions enhance many presentations. Visitors with hearing difficulties or limited language skills particularly may depend on the combination of written and spoken words with visual images, and all visitors are likely to benefit from systems which engage several of the senses at one time. However, the addition of audio can create real environmental problems.

Listening booths may be an attractive option for visitors who want the privacy to experiment with the system or concentrate on detailed information; however, booths take up space, encourage long visits, and do not invite casual interaction or observation by passers-by. Headphones or telephone-style handsets also offer some privacy, again at the expense of the those who cannot vicariously engage with the presentation.

Observation of headphones and handsets in, for example, St. Paul's Cathedral or the Pompidou Centre, suggests that many visitors are shy of handling things which they – rightly – think they could break. Hoods or shields, built into the housing or immediately around it, offer a compromise between the privacy of a listening booth and the easy access of any other gallery exhibit. Equally, many institutions alternate silent and audio terminals to maximize the distance between their speakers.

The National Gallery of Canada in Ottawa uses audio in a display which explains how artists train. Text and pictures are complemented by a series of readings from diaries and letters, which run on a continuous loop. Visitors sit on padded benches, and listen with headphones. The choice of language and volume control are both effected through a sliding bar mounted by each seat: moving the control to the largest of four blue dots produces maximum volume in English, while sliding it in the opposite direction to the largest of four red dots produces maximum volume in French. Bi-lingual signage provides the simple instructions for use and identifies the source of each reading.

A cheap lesson from the point-of-sale market notes that staff are more than likely to turn off the sound altogether if the continuous attractor loop repeats the same spoken invitation or jingle many times an hour. Most such systems employ a silent attractor loop, and many are silent altogether.

The second part of this report includes several designers' views on housings, and some observations of visitors' reactions to them.

TECHNICAL SUPPORT

No matter what the eventual media mix, technical support is essential to maintain displays from day to day, and keep abreast of new developments as systems are enhanced and exhibits change.

It is likely that technical support will be sought on two levels. There is usually a need both for a versatile and imaginative all-rounder with a high level of diagnostic and psycho-motor skills (the techie), and a robust and patient soul who will spend a great deal of time moving and mending awkward and/or heavy equipment, often in confined and poorly-lit spaces (the roadie). Neither position stands on a long career path, and both offer considerable scope for hard work under acute pressure with few spiritual or material rewards.

The experience of most production companies suggests that it takes a distinctive personality to fill either role, and turnover is high in both. The multimedia products of the early 1990s are much more reliable than the IV systems of the mid-1980s, and serious technical problems are more rare. Nonetheless, in-house technical support is vital where a variety of systems employ different combinations of hardware and software, and are expected to deliver a reliable service to the public every day.

However, it is only just a joke that 95% of system faults result from 'failing to turn it on'. One supplier reports that, after a year's contract with a national telecoms group, fully 75% of reported faults were due to operator error, and the balance to defective equipment from the manufacturer.

It is easy even for experienced operators to overlook one switch in a system with several separate controls. A single power source for each system, gallery or building effectively automates daily start-up and shut-down procedures. The use of timers also helps, although readers are advised to consider the reports cited here from the British Golf Museum, the US National Museum of Racing, and the V&A.

In an IV system, it is also easy to overlook one of several software sources, or put the wrong one in the right drive. Trainers often have to change AV media and computer disks several times a day, but a gallery should be able to leave software in the system for the duration of an exhibition. Equally, many systems now include a diagnostic routine which ensures that all the complementary software is installed and running properly before the programme begins – or sends out a clear error message to indicate what is missing or malfunctioning.

Any new project is vulnerable to teething troubles, but everyday problems are most likely to originate from global factors such as power surges. Of course, the hardware must be sturdy and tamper-proof, with software designed to ignore unauthorized activity such as gratuitous key presses or touches to the wrong part of the screen. Simple conventions consistently employed help the visitor to use the system efficiently, and reduce the chance of inadvertent, malicious or experimental interference.

Systems should return to attractor loops if left unattended when, for example, a visitor loses interest and walks away. Of course, the system should first warn the user and offer alternative action for those who may simply be slow, or discussing what they have found with others in front of the screen, so that the programme does not simply disappear.

The combination of system design and programme design should anticipate many simple operational problems. At the same time, everyone who is likely to work with or near the systems – custodial as well as curatorial staff – should be introduced to the hardware and programmes. Visitors may need encouragement to engage with the systems for the first time, and the staff actually in the gallery are ideally placed to help. (It is also observable that many custodians enjoy telling the visitors about *their* room: when they are trained to do this well, visitors are often engaged by their interest and goodwill. Conversely, indifference or hostility from the staff has scuppered many IV initiatives before now, in environments of all kinds.) A short internal training programme with a strong motivational message can accomplish much.

The value of maintenance contracts for systems which are out of warranty is difficult to calculate: neither the AV nor computer industries as a whole are over-supplied with service engineers, and interactive multimedia systems particularly require a special order of expertise.

Many suppliers offer maintenance contracts, but some institutions have found these to be no more than an expensive form of insurance. Telephone hotlines may be cheaper, especially where in-house staff only lack experience of fault-finding on the new system. Typically, day-to-day maintenance can be achieved quickly and economically in-house, with a provision in the budget for calling on external support or returning faulty equipment to its manufacturer when the problem is serious.

DESKTOP VIDEO

Digital audio and video products provide a variety of tools which put creative control into the hands of programme-makers, and even end-users, who need only a suitably-equipped desktop computer to perform tasks once unique to professional AV production facilities.

Typically, the digital production suite will include cameras and scanners to capture film, video, slides and other visual material such as computer graphics, plus the computer hardware and software to refine and edit these images on screen, and create complementary computer-generated images, text and animation. Some include recording and editing facilities for speech and music, and even create music and sound effects electronically.

In practice, most commercial developers use a combination of in-house resources and external facilities to process the variety of material which a typical multimedia production may incorporate. However, at a more modest level, a new generation of products for desktop computers allows end-users to create effective multimedia packages in the office or classroom, or to maintain a presentation created by professional consultants.

Conventional video production demands that special effects be executed in an editing suite, and recorded immutably on a master videotape (on which, in IV, the master videodisc is based). Starting with the same undistorted image (stored on videotape or disc, or computer disk), digital video adaptors produce many of the same effects entirely within the delivery system, by reprocessing the data in the video signal as the programme is presented.

These products are typically supplied as add-on boards, which plug into the body of the computer, and/or computer software. With their processing power, data which would ordinarily comprise a single full-screen image can be converted to a digital format and manipulated to create any number of effects in various places within the interactive presentation. The original remains intact on the source medium, to be presented in whole or in part as the control program dictates.

Thus, in 1990, the multimedia centre at Leicester Royal Infirmary used a videodisc made some years before to demonstrate new presentation techniques. With a DVA4000 board from VideoLogic and Microsoft Windows 3.0 software, two separate stills from the disc can be presented in adjacent windows on the screen, together with text which describes the similarities or differences between them. The images may be cropped and/or reduced to fit the window, but the unadulterated image can also be displayed across the full screen in an instant. These options are held not on the videodisc, but in the computer program which controls the presentation.

JPEG & MPEG

The Joint Photographic Experts Group (JPEG) and Motion Picture Experts Group (MPEG) are international consortia of hardware, software and publishing interests dedicated to developing international standards for the compression of, respectively, still and moving video images in digital environments such as computers and compact discs.

Even before their formal adoption by the International Standards Organization (ISO), the recommendations of the MPEG and JPEG committees are creating *de facto* standards to which many manufacturers will conform in the design of new multimedia products. Thus, the two acronyms are often used, as adjectives and nouns, to describe products which should meet international standards in digital video production.

A number of manufacturers, from large corporations such as the Japanese Victor Company (JVC) to influential developers such as C-Cube, Iterated Systems and VideoLogic, are developing both their own compression/decompression algorithms, and hardware and software products based on the MPEG and JPEG recommendations.

This market is developing rapidly, and in many directions. One line of innovation has already adapted the JPEG recommendations to moving footage, as 'moving JPEG'; another, from Iterated Systems, employs the power of complex formulæ called fractals (which describe irregular shapes and textures from the natural world, rather than the ideal forms of Euclidian geometry) to analyse and compress the data in AV signals. Products now available from various manufacturers offer compression ratios of anything from 10:1 to 100:1 or more, with a trade-off between higher image quality and higher compression ratios. Some systems already demonstrated ultimately promise compressions to factors of 10,000 and even 20,000 – sufficient to send AV signals over ordinary telephone lines.

This is effectively a watching brief for anyone seriously interested in new technology. Professionals in design and marketing environments should understand these principles and be aware of the buzzwords. However, designers who want to create an interactive multimedia presentation for immediate use and enjoyment can confidently employ platforms and products available now, and follow a gentle upgrade path as new developments are made practicable. To stay on the learning curve, it is useful to be aware of trends which will influence the hardware and software markets of the future (typically, through reading the specialist press and visiting trade shows and exhibitions). A firm foundation – informed by analysis of the institution's own needs and resources, and the advice of specialists in the multimedia market – should accommodate practical innovations within this volatile market.

DESKTOP VIDEO PRODUCTS

Many suppliers, including leading hardware manufacturers and software developers, now offer products which allow specialists in fields other than computing to create multimedia presentations entirely on the desktop. The 1992 *European Multimedia Yearbook* lists over 400 products for professional developers and end-users, for every stage of production from image capture through software design to integration of the finished package.

IBM's Audio Visual Connection (AVC), for example, is a presentation system which can be used to design and deliver interactive programmes combining text, data, sound and pictures; IBM's M-Motion Video Adaptor handles up to three video sources (including a camera) as well as graphics and sound from the computer disk. Apple's QuickTime is a comprehensive toolkit which allows designers to incorporate still and moving pictures, and sound, into virtually any Macintosh software package. With such tools, a word processing document could literally include a video clip, to illustrate what a thousand words could not describe.

Desktop video demands fast, powerful computers with plenty of storage, and the equipment to capture conventional AV material as digital data. This demands investment in equipment and expertise but, in the hands of good designers and skilled operators, it represents an opportunity for creative control and cost-effective in-house production.

DIGITAL VIDEO INTERACTIVE (DVI®) TECHNOLOGY

Another approach to the demand for video in a desktop computer was developed initially by RCA and the General Electric DVI Technology Venture and, since its acquisition in 1989, by Intel. DVI technology supports a versatile presentation in a digital environment, but its principles are quite different from those of other multimedia platforms.

Consumer products such as CD-I and CDTV are based on CD-ROM (that is, a physical storage medium); DVI technology comprises a set of computer chips which use mathematical algorithms to analyse, compress and store digital video data and reconstitute these images subsequently.

DVI is thus a process as much as a product, an adjective rather than a noun. The storage medium could be CD-ROM, computer disk or WORM media. The chips are now mounted on a set of plug-in boards, which will soon be reduced to a single board, and incorporated into the computer's motherboard. The technology was developed for the IBM PC environment, but has been adapted for the Apple Macintosh. Built into the heart of the machine, DVI technology may become a standard feature of a new generation of multimedia computers.

DVI has overcome the problems associated with high levels of data storage and transfer rates in digital video technology by developing very efficient compression techniques. The Picture Image Compression (PIC) algorithm is based on the JPEG standard. Realtime Video (RTV) supports video compression on the user's own computer. Production Level Video (PLV) must be compressed at a professional production facility but provides video of good visual quality which can be decompressed on the user's computer. Subsequent enhancements of the DVI chipset will incorporate new industry-standard algorithms including JPEG and MPEG.

DVI technology supports multiple audio tracks, high resolution stills (768x480), graphics, full screen motion video at medium resolution (256x240), and features such as texture mapping over 3D objects, and panoramic views processed from stills.

Storage capacity depends on the choice of medium and data in a given application – with the typical 650MB storage capacity of a CD-ROM disc, DVI would typically support 72 minutes of moving video with audio, or 20 minutes of video with seven hours of separate commentary, and 5000 or more high resolution still frames. Ultimately, of course, the specific choice of sound and pictures in any one presentation determines how the available storage and processing capacity is employed.

DVI technology currently supports networks – that is, several terminals linked together, sharing information and resources – better than other multimedia platforms. This may be an advantage in tailoring presentations for use in different parts of the same site, or by different audiences, in corporate environments and gallery displays.

The European DVI Developers' Group provides a communications and support network with a newsletter, directory, workshops, meetings and bulletins about DVI technology and potential markets. Contact Isobel Pring, Digital Vision International Ltd., Suite 202, The Foundry, 156 Blackfriars Road, London SE1 8EN (UK), Tel. +44(0)71 721 7053, Fax +44(0)71 721 7054.

OPTICAL DISCS & VIDEODISCS

Some common principles apply to disc systems from 30cm/12" LaserDisc through the 12cm/4.75" CD in all its forms, and mini-discs as small as 5cm/2" in diameter. Apart from size, the principal difference between incompatible formats lies in the type of data which they hold, and the way these are recorded. Nonetheless, reflective optical discs of all kinds are manufactured and played in essentially the same way.

All data (both AV and computer signals) are recorded in a spiral channel patterned to record the binary computer code common to all digital media. That basic choice of two signals – 0/1, on/off, magnetic/unmagnetic – holds the key to an extremely accurate and efficient means of data storage.

Long strings of binary code, peculiar as they may seem to the human mind, can be retrieved, processed and transmitted at phenomenally high speeds by electronic devices. The margin for error is much lower, and the facility for error detection and correction much higher, with the rigid codes of digital data than with the constantly fluctuating signals of analogue media such as videotape or LP records.

In disc production, the digital pattern of pits and shallows is cut into a plastic master disc with a high-powered laser beam. From this, sturdy metal stampers can be made to replicate many copies at a professional pressing facility; alternatively, single copies or low-volume runs can be cut in 'real time' – that is, individually, at the same speed at which they would normally play – at a special facility or, with the right equipment, in-house.

TRANSMISSIVE AND REFLECTIVE

Transmissive optical discs are translucent; reflective optical discs cover the pattern of pits and shallows with a thin metallic film – silvery aluminium distinguishes discs which are pre-recorded from the golden sheen of those on which users can store their own data. Both seal that surface under a protective plastic topcoat, which only the strong cutting laser of the recordable systems can penetrate.

Optical systems retrieve data from the disc by shining a low-powered laser beam onto its surface. Transmissive systems shine the beam through the disc to a reading head; reflective systems bounce the light off the metallic film through a network of prisms and mirrors within the disc player to a photo-diode which decodes variations in the reflected light back into audio, video and computer signals. These may then be converted back into analogue signals for distribution to the output devices such as conventional TV or hi-fi systems.

SALIENT FEATURES OF OPTICAL DISC

The combination of digital data storage and the laser beam means that all optical discs formats share some important characteristics:

- There is no physical contact between the recording and the reading device. Thus, constant use, even of a freeze frame held motionless on the screen, will not wear out a disc as it will a videotape.
- The transparent plastic topcoat protects the recorded information from dust, scratches and accidental damage, to which videotape, slides and floppy disks are all much more vulnerable.
- Ordinary optical discs, once recorded, cannot be changed or amended. In an interactive system, separate computer software may restrict access to outdated material or supply new data in the form of computer-generated text or graphics, but the only way to change the disc is to amend the master tape and start again.
- Digital data from the disc can be manipulated in various ways within a computer-based system so that, if the contents of the disc are still sound, the medium itself can often be re-used with new hardware and software as delivery systems evolve and upgrade.

Optical discs are less vulnerable to damage, dirt and wear than computer disks or videotape, and their players are proving more robust as well. LaserDiscs have been reported to survive an afternoon in the back window of a car parked in Arizona in the summer, and a night in a player in an unheated exhibition pavilion in Tokyo in the winter. However, some have also been destroyed in transit by the UK Post Office; although they are generally more resilient than other AV or computer media, optical discs should be treated as carefully as audio LPs.

Optical discs are extremely difficult to replicate illegally; unfortunately, a single LaserDisc provides an excellent source for any number of pirate videotapes, but manufacturers are looking at anti-copy chips, encryption codes and other technological solutions to prohibit this. For the producers of interactive multimedia programmes, optical discs offer a recording medium which has so far proved secure from piracy.

Some organizations have also found that optical disc players are less vulnerable to theft or misuse because there are as yet few consumer entertainment discs on the market to support clandestine use.

VIDEODISC FORMATS

Most commercial optical discs, including all LaserDisc and CD formats, are reflective. However, the American aerospace manufacturer McDonnell Douglas developed a transmissive disc which has enjoyed some use in professional applications, and Compact Video Disc (CVD) is an analogue/digital hybrid developed by SOCS Research.

In fact, the concept of the videodisc goes back to John Logie Baird, who produced a videodisc for the home gramophone in the 1920s, and some dozens of systems have been patented (and even released) over the years. Only a handful have had any market presence, and two of those – RCA's Capacitive Electronic Disc (CED), SelectaVision, and JVC's Video High Density (VHD) – cost their backers a lot of money before withdrawing from the field in the mid-1980s. The subsequent emergence of compact disc systems has largely discouraged any further attempts to launch new videodisc formats, so most videodiscs are now LaserDiscs.

LASERDISC & LASERVISION

The most successful videodisc format was developed by Philips, supported by Sony, Pioneer and many other large manufacturers of home entertainment products. LaserDisc is particularly popular in Japan, where discs outsell pre-recorded videotapes on the consumer market. With a relaunch headed by Pioneer, 10 years after its first release by Philips, LaserDisc is now gaining popularity in American and European markets as a high-quality medium for movies and music videos.

A 30cm/12" LaserDisc can hold at least 55,000 single images on each side – enough for 30 minutes of moving footage in the NTSC video standard or 36 minutes in PAL/SECAM (explained on page 13). The disc also has two audio tracks, which can be recorded together in stereo or separately for more audio capacity. (The principal difference between the original LaserVision and the current LaserDisc format is the enhancement of the audio facilities to digital quality.) LaserDisc can also hold computer data such as instructions to the player's own microprocessor, computer control programs, and computer-generated text and graphics.

In 1987, Philips launched the CD-Video (CD-V) format, as a 12cm/5" single for a six-minute music video with another 20 minutes of CD-DA audio, and a 20cm/8" format with 20 minutes of video. However, combi-players – which handle CD-DA and LaserDiscs of various sizes in a single machine – entered the American market around the same time, and soon proved much the most popular option with consumers. In November 1989, Pioneer released its trademarked LaserDisc name and logo to the market at large and, in February 1990, Philips adopted both for its videodisc products. Thus, all reflective optical videodiscs are now effectively LaserDiscs, and many people use the terms LaserVision, LaserDisc and videodisc interchangeably.

The European LaserDisc Association (ELDA) aims to promote knowledge of the LaserDisc system, development of applications, standardization and interchangeability, and use of the LaserDisc logo as identification of compatibility. Contact ELDA, Kastanjrlaan 7, Building SFHS, PO Box 80002, Eindhoven 5600 JB (Netherlands), Tel. +31(0)40 736094, Fax +31(0)40 733512.

RECORDABLE VIDEODISCS

Although most discs, like most vinyl records, are designed for pre-recording, some are recordable. These are particularly useful in archiving, but sometimes offer a cost-effective medium for single copies or low-volume runs in other applications.

Write Once-Read Many (WORM) drives are principally used as computer peripherals for mass data storage; once recorded, they cannot be erased or rewritten. This can be an asset in an archive where data must be secure from accidental erasure, or alteration of any kind. Typically, a 30cm/12" disc has a capacity of at least one gigabyte per side (*i.e.* half again as much as a 650MB CD-ROM), but storage capacities vary in products from different manufacturers. WORM systems are excellent for document storage, less useful for display work.

Manufacturers including Panasonic, Sony and Teac have systems for the in-house recording and replication of videodiscs. These WORM systems store analogue AV rather than digital data. The catch is that each has its own format and hardware (incompatible with LaserDisc and each other).

These systems are still expensive to buy, although the hardware itself (or at least its use on someone else's premises) can sometimes be hired. The need for in-house recording of this kind tends to be limited to areas where confidentiality is paramount, or where the facility to capture all manner of media archivally, or build an image base gradually, justifies the investment in hardware and operating skills.

Erasable discs go a stage further in that, once recorded, they can be erased and re-recorded. There are currently three erasable optical technologies: magneto-optic, phase change, and light-sensitive dyes. All are in early stages of development, and it is by no means certain which will ultimately succeed. The most obvious applications for erasable optical technology are in computing, as an alternative to mass storage on magnetic hard disks.

COMPACT DISCS

Initially, CDs for the consumer market were designed for use with the TV and the hi-fi, and professional formats for use within computer-based systems. Multimedia products are now bringing CDs to the home computer, and inexpensive new platforms to professional environments. The 1992 edition of TFPL's *CD-ROM Directory* lists 2012 software titles in CD-ROM and XA, CD-I, CDTV and DVI formats – without considering the thousands of audio discs on the most familiar format, CD-DA.

COMPACT DISC-DIGITAL AUDIO (CD-DA)

CD-DA is the format which most people now recognize as *the* compact disc. Its audio quality, storage capacity and durability are widely perceived as superior to those of other audio systems. The players are relatively cheap and, if the discs are pricey, at least they should last for years. Common principles of upward compatibility mean that a CD-DA disc can be played in a CD-ROM/XA, CD-I or CDTV system.

The main market for CD-DA is entertainment, but the format also offers a simple and cost-effective means of supplying audio to galleries, enhancing computer text and graphics or providing extra audio to other media. CD-DA jukeboxes provide compact and reliable public entertainment services.

A variation called CD Plus Graphics (CD+G) uses the disc's spare data channels to record text and images which can be decoded in an appropriate player to display notes, lyrics, pictures and other background information on the TV screen while the audio plays from the hi-fi. Some music discs employ this simple multimedia capability, but few consumers yet have CD+G or compatible players.

For the British Library, CD represents an ideal medium to store and distribute material from the National Sound Archive. CD-DA typically supports around 72 minutes of stereo, but new techniques are stretching the capacities of all optical disc formats, and other CD platforms trade higher capacity for lower quality.

COMPACT DISC-READ ONLY MEMORY (CD-ROM)

CD-ROM is a small, robust medium which can record as much as 675MB of data, text, graphics, digitized audio and images (again, this is an upper limit – 650MB is typical). With storage capacity equivalent to some 250,000 printed pages, CD-ROM is widely used in professional markets for databases and archives, both published and private.

Some critical standards for the use of CD-ROM were developed after the product itself – ISO 9660, High Sierra, Rock Ridge and CD-RDx – but there is widespread compatibility for CD-ROM within the separate orbits of the IBM PC and Apple Macintosh. This rarely affects the designers of exhibitions unless material is developed with a view to distribution on CD-ROM to educational or other software markets. It is then important to seek professional advice from a disc presser such as Nimbus, Philips or Telemedia/Sonopress in preparing data for disc.

CD-ROM EXTENDED ARCHITECTURE (CD-ROM XA)

CD-ROM has limitations in the delivery of fully integrated multimedia presentations, which the extended architecture of the XA platform addresses. Unlike CD-ROM *per se*, CD-ROM XA has universal standards drawn from the CD-I specification for encoding audio, graphics and still pictures on CD.

CD-ROM stores and processes text, graphics and audio in separate blocks, which greatly impedes the flow of AV signals from the disc through the processor to the screen and speakers. CD-ROM XA interleaves data of different types to facilitate faster and smoother presentations. The rapid processing of graphics can effectively simulate simple animation in CD-ROM XA. The XA format supports two of CD-I's four image encoding formats, two of its three audio formats, and CD-DA audio.

XA also supports both CD-I's unique CD-Real Time Operating System (CD-RTOS) *and* the industry-standard Microsoft Disk Operating System (MS-DOS). Thus, an XA disc could be used in the PC at work, and the CD-I player at home – an interesting crossover market for software of all kinds, including training, education and reference material.

CD-INTERACTIVE (CD-I)

CD-I was launched by Philips on the US consumer market as 'Television with a mind of its own – yours'.

CD-I supports up to one million words of text, 6340 photographs or graphics, 2.4 hours of LP-quality stereo, 4.8 hours of FM-quality stereo, 9.6 hours of AM-quality stereo (and twice those capacities in mono), 650MB of computer data or – in practice – a combination of still and moving footage, computer text and graphics, animation, music, speech or sound effects. The system has some capacity to store volatile information, so gamesters can record best scores or running tallies even if the player is turned off between games. CD-I comes in one box, with a remote control device that combines a tiny joystick (thumbstick), buttons and a keypad. The system has been used with touchscreens and trackerballs in museums in the UK and US, and Sony has demonstrated a portable version.

CD-I trades quality for capacity like any other multimedia platform, with three audio formats (each in stereo or mono) and four graphics formats, as well as CD-DA audio and video. Accommodating this trade-off is one of the principal challenges facing the designers of multimedia software in all platforms. Fortunately, CD-I's specification – prepared by Philips with support from Sony, Matsushita and others – ensures technical compatibility between all CD-I discs and players.

CD-I was announced in the spring of 1986, and launched in the US in October 1991 and in Europe in April 1992. It is usual to announce products well before their commercial release, although this gestation proved rather longer than most, and occasioned some sceptical comment from observers who had little else to discuss in the interim. However, the $1000 launch price (actually, $800 in most shops) compares favourably with that of CD-DA, as does the launch catalogue of some 50 titles, with hundreds more titles in production.

These range from storybooks through teach-yourself packages to encyclopædias. As a consumer product (and, in time, an inexpensive resource for education), CD-I could open a new world of publishing which museums, galleries and libraries are ideally placed to exploit; the Smithsonian Institution and the British Golf Museum have already contributed to CD-I discs based on their collections.

The European CD-I Association provides services and opportunities for everyone involved in the use and production of CD-I. Contact Julian Lynn-Evans, Mullard House, 1-19 Torrington Place, London WC1E 7HD (UK), Tel. +44(0)71 331 1569, Fax +44(0)71 580 6757.

COMMODORE DYNAMIC TOTAL VISION (CDTV)

Commodore launched CDTV in consumer markets in the US and Europe in April 1991, with $1000/£600 players which combine features of their Amiga computer with a CD-ROM drive. An upgrade kit addresses a captive market of some two million Amigoids, dedicated games players (who, in its first year, seem to be the product's most receptive market).

Both CD-I and CDTV players connect to the TV, hi-fi or a MIDI instrument such as an electronic keyboard, and both accept CD-DA discs. As well as the basic keypad, CDTV offers input devices including a joystick and keyboard, and smart cards to record data such as scores in a game. CDTV supports high-quality computer graphics, animation and audio, with a storage capacity of 540MB.

By the autumn of 1991, Commodore reported that some 380 developers worldwide were working on CDTV titles, and over 200 would be in circulation within the year. The consumer catalogue, with most prices between £30 and £45 per disc, features children's entertainment and games of all kinds, with some teach-yourself and reference material. Commodore has also worked with Global Learning Systems and Derbyshire County Council to produce a Japanese language programme for schools (inspired by Toyota's £700m car assembly plant in that county).

PHOTO CD

Kodak's Photo CD is to still images what CD-DA is to audio, trading relatively low volume for extremely high quality. Photo CD holds up to 100 images from ordinary 35mm film at a range of quality levels up to four times that of HDTV. With appropriate players, these can be displayed and manipulated on the TV, or in the computers of desktop publishing or AV production facilities. Kodak says that Photo CD will also soon handle audio which consumers can record at home, and video images.

For a premium of $20/£10, consumers can have both prints and the Photo CD disc. This can be built up over time (say, with four rolls of 24 shots each). Discs can be copied, and users can assemble their own 35mm images for transfer to disc. When the product is released in the US in mid-1992, dedicated players will cost from $399 to $599; however, the format is compatible with CD–I and CDTV systems. Kodak promises some 200,000 'drop points' for film in the US; the processing unit costs around $100,000.

Kodak estimates that some 250 million 35mm cameras produce 60 billion images a year. The potential for publishing on Photo CD ranges from archival and reference collections to resources for professional designers – markets which galleries, museums and libraries are ideally placed to exploit.

FULL-MOTION VIDEO (FMV)

Both CD-I and CDTV were launched with players which can be upgraded to full-motion, full-screen video (FMV) but are initially restricted to video in windows, often at reduced speed and resolution.

This is because FMV simply demands too much storage space and processing capacity for the first generation of consumer players. Its introduction thus depends in part on the quest for inexpensive, universally compatible compression systems, and specifically on the work of MPEG and JPEG.

CD-I could already demonstrate FMV at the time of its European launch. However, although true video will certainly enhance these products, multimedia by definition will continue to use a variety of resources strategically to meet the needs of each programme and the demands of each platform.

VIDEOTAPE AND SLIDES

The British Library, like some museums and galleries, has its own in-house video production facilities as well as a large collection of still images on 35mm slides and other photographic media. Tapes and slides support part of the Library's education service, with some two dozen video or tape/slide shows in a typical month.

Tapes and slides have distinct advantages in this environment. There is little resource material available off-the-shelf in any optical disc format, but both slides and videotapes can easily be made, hired or purchased, often at short notice. Many people like to maintain their own slide collections, and this is a pocket of resistance which should not be underestimated in designing centralized systems of any kind. Equally, tapes and slides are both easily supported by in-house production services. It is likely that slides and tape will have a role in the education service at least for some time to come.

DISC VS. TAPE

Tape is essentially a linear medium, and it is difficult to support even simple random access when the tape has to wind back and forth every time it moves from one segment to another. Uniquely identifying every single frame of the recording is more difficult on tape than disc and this, in turn, reduces videotape's effectiveness in achieving the frame-accurate retrieval which disc effortlessly supports – and which is vital to interactivity.

Furthermore, all data are recorded on the exposed surface of a video or audio tape, which deteriorates slightly every time it passes through the video and audio heads which read it. Tape is consequently more vulnerable to wear and tear than disc, which protects its recorded surface with a clear plastic topcoat.

For much the same reason, it is more difficult to hold still frames on tape than disc: the reading head of the videotape player constantly wears away the surface of the tape, while there is no physical contact between the reading device and source medium in a disc player. Some new videotape players are designed to meet these criticisms, but a professional-quality machine of this calibre is at least as expensive to buy as a videodisc player, and usually much more expensive to maintain.

There are many excellent arguments for employing LaserDisc and CD exclusively in exhibitions, and thinking seriously about their use in education and other services. It is now relatively simple and inexpensive to produce discs of various formats singly or in small numbers, particularly to create archival resources which can be adapted to any number of applications with appropriate programs to select and arrange images for display. A new range of software products provides these tools for people with little or no experience of computer-based design. In the longer term, a good set of resource discs, and a user-friendly authoring package, could allow lecturers, researchers and even museum visitors to create their own presentations from slide collections on disc.

ANECDOTAL EVIDENCE

Several organizations now use videodisc exclusively – for linear as well as interactive presentations – simply because the discs and players have proved much more reliable than videotape systems.

In 1988, the Science Museum in London calculated that a single videotape system cost £2420 to set up, compared to £2195 for a LaserDisc system – that is, including capital costs of players and controllers, and maintenance costs for service contracts, parts and consumables. Even allowing £3000 for disc production (and no comparable expense for videotape), after 18 months, they calculated that two installations, of four systems apiece, would have cost £10,320 for tape and £9530 for disc.

Guinness World of Records' London exhibition coordinator, Graham Hewison, cited the high cost of replacement heads for its U-matic videotape systems in justifying the move to videodisc; Guinness exhibitions in other cities also deliver video images on disc.

In the mid-1980s, when the Cecil Gee men's shops added fashion videos to its shop displays, it proved cheaper to produce one videodisc than to replace worn videotapes several times a season.

The Winchester Arms Museum in Cody, Wyoming, converted its 13-minute multi-image slide show to videodisc in 1985. The programme runs some 3000 times a year – four times an hour during the height of the tourist season. Three years later, curator Herbert Houze reported to *The Videodisc Monitor* that 'the system has had almost no down time. In fact, the back-up has never been used. Because of the savings in equipment repair and replacement costs, the cost of the conversion paid for itself in less than a year.'

HOLOGRAMS

Over the past two decades, holograms have developed into a range of specialized products with applications in several disparate fields. The basic principle remains the same for all: an image recorded on a specially processed film, when properly illuminated and viewed from the correct vantage point, creates the ethereal impression of a three-dimensional object floating in space.

Coherent light – specifically, a laser – is vital to the production and projection of holographic images, and companies which produce holograms often deal in laser light shows and related display services.

In trade shows and exhibitions, holograms are among the upmarket options for attracting attention in a competitive environment. In a noisy hall, the fact that they do not carry sound can be an advantage. A 1.5mx1m version of the Michelin tyre man, Bibendum, was perhaps the world's largest (and heaviest) hologram. It was made by Holographics (UK), who also make holograms as small as 25x20cm, and use laser graphics to reproduce logos, text, still and moving pictures, and 3D images.

The holographic industry has cultivated another market in recent years with the incorporation of holograms into the security systems such as those used on plastic ATM and credit cards.

There is a small but lucrative market for holographic art, and related novelties such as stickers, badges and jewellery. The Light Fantastic Gallery in London's Covent Garden Market provides a showroom for holographic images and novelties from small images in large editions which sell for as little as £10 apiece, to limited editions which can fetch thousands of pounds.

Portraits of photogenic pop stars such as Michael Jackson are popular in the upper price range, and there is a lively market for the holographic equivalent of sporting prints: Neil Irwin of Light Fantastic reported that a limited edition of four falcons (a popular subject among collectors in the Middle East) doubled in value within a year of its release. Limited editions generally run to 50 or 100 copies, and popular images will sell out within six months.

The interactive potential of holograms is slight, although Light Fantastic has featured a display which allows visitors to shake a spectral hand (an experience of sufficient novelty to compensate for its limited educational or æsthetic value). However, for exhibitions and displays, holograms do have some particular attractions.

Holographic portraits can be extremely striking, and eerily realistic. Where the familiar devices of portrait busts or coloured pictures hardly attract a second glance, holographic images can be riveting. The ill-fated Royal Britain exhibition, described on page 63, used holograms to animate the face of a life-size model of Elizabeth I delivering the Tilbury address. In the UK, the National Westminster Bank's ATM Servicecard features a tiny portrait of Shakespeare – who smiles at the viewer as the card is turned from right to left.

Holograms can also create haunting reconstructions of images distorted or destroyed by time. Light Fantastic has displayed holograms of a Saxon helmet and the reconstructed head of Lindow Man, and mounted a broken goblet in front of a holographic image of the complete vessel so that the object can be viewed on its own, or against its translucent reconstruction. The British Museum itself displays a hologram of Lindow Man's desiccated body (made by Richmond Holographic Studios), adjacent to a picture of his reconstructed face, and the body from the bog itself.

The cost of producing a hologram varies dramatically according to the size, colour and quantity of images required; the process certainly demands the skills and facilities of a professional supplier. The production process begins by making and photographing a three-dimensional object or model. A typical production schedule requires up to three weeks for artwork and modelling, two to four weeks for mastering, and two weeks for production.

Mastering typically starts around £1500 for a tiny image, and may run to £30,000 for something the size of Bibendum. A run of fewer than five copies may cost £1500 each, but large runs may be as low as £10 or £20 per image. Effects such as multiple exposures (for an image which appears to move as the viewer walks past it), or extra colour, add to the production cost.

Rarely do actual gallery displays lend themselves directly to commercial exploitation. However, with the select market for holographic art, there is an unique commercial opportunity to subsidize the cost of producing holograms for display, by publishing copies as a limited edition which allows collectors to buy the very image that they saw in the museum or gallery.

HIGH DEFINITION TELEVISION

High definition describes television and video systems with a level of clarity and detail hitherto seen only on film. There are as yet no international standards for high definition TV (HDTV) or video (HDV), nor much non-broadcast production experience outside Japan. The potential here is great, but largely unrealized.

HDTV STANDARDS

The two prominent HDTV standards are MUSE and HD-MAC. Both double current standards for horizontal and vertical image definition: MUSE, developed in Japan where 525-line NTSC prevails, employs 1125 lines per screen; HD-MAC, developed in Europe where 625-line PAL/SECAM prevail, employs 1250. (The prevailing international television standards – NTSC, PAL and SECAM – are explained on page 13.)

Its promoters believe that HD-MAC offers superior performance, and supports both terrestrial and satellite services better than does MUSE. Both expand the width/height aspect ratio of the screen from 4:3 to the 16:9 'letterbox' format.

Most significantly, both are based on analogue compression systems which demand powerful processors in the HDTV decoder. To achieve this in a reasonably-priced TV set demands even smaller and faster integrated circuits than the semiconductor industry has yet developed. Thus, no HDTV system is ready for a mass market; in the meantime, the emergence of new satellite television services and digital video compression systems could seriously undermine present HDTV initiatives.

THE STRUGGLE FOR HDTV

Since 1986, European manufacturers and broadcasters have blocked Japanese attempts to establish their MUSE high-definition system as a world standard. A European consortium headed by Philips, Thomson and Bosch is promoting the HD-MAC system, which was developed through the European Community (EC) Eureka Project 95. This £150m initiative united governments, broadcasters and manufacturers to develop an HDTV specification suited to prevailing European standards.

In 1990, 14 electronics companies, programme-makers and broadcasters set up Vision 1250 as one of the first European Economic Interest Groupings. Vision 1250's declared aim is to 'set up and manage both fixed and mobile HDTV facilities to be made available to TV and audio-visual professionals and broadcasters in the European Community'. Public demonstrations of HDTV commenced in July 1990; prominent show pieces include transmissions from the 1990 World Cup, 1992 Olympic Games and 1992 World's Fair.

SATELLITE BROADCASTING

The future of high-definition also depends on another initiative, for the Eureka HDTV consortium based the HD-MAC standard on the D2-MAC standard for satellite broadcasting.

The Multiplexed Analogue Component (MAC) system was developed in the UK and adopted by the European Broadcasting Union in the early 1980s, under the influence of views formed in the 1970s that the future of satellite broadcasting lay in powerful Direct Broadcast Satellites (DBS).

In the event, less powerful satellites, which offered smaller and less conspicuous receivers (typically, 60cm/24" dishes), proved more popular with consumers. These were not subject to the EC's directives for DBS, and preferred to use prevailing PAL broadcast standards rather than D2-MAC. The issue was purely practical: D2-MAC demanded enormous investments in new hardware (and skills) by programme-makers, broadcasters and consumers, all of whom were already equipped for PAL.

At the heart of the HDTV wars lies this conflict between manufacturers eager to launch a new range of products, and suppliers and consumers who need strong incentives to replace their present TV and video systems with something entirely new.

THE HARDWARE MARKET

The market for AV products in the industrialized world is saturated – only something radically new and different is likely to create a fresh demand for the vast range of hardware needed to produce and disseminate television and video to corporate and consumer markets.

To receive HDTV or satellite broadcasts, viewers need suitably-equipped TVs. Some new sets can receive standard broadcasts and MAC satellite signals in the 625-line PAL standard; a small 'black box' can also decode MAC signals for display on an ordinary TV. Any set equipped to intercept MAC can also pick up HD-MAC, although the picture will of course lose its high definition on an ordinary television screen. To receive MAC satellite broadcasts in high definition letterbox formats requires another TV altogether, but little so far has encouraged consumers to want HD-MAC enough to buy a new set.

Backward compatibility may help to encourage programme-makers to work in HD formats which can be viewed on ordinary sets now and new systems in the future – a strategy employed in the early days of colour TV. Popular programmes could thus win ratings, advertising and funding now, and enjoy a long shelf life. The MUSE system does not now support backward compatibility: only those with brand-new sets can receive MUSE broadcasts.

HDTV IN THE US

The Federal Communications Commission (FCC) will not decide on an HDTV format for the US until 1993, but has indicated that it favours a system with an upgrade path over one which demands entirely new hardware. On 4 August 1990, *The Economist* reported that NBC, America's largest network, 'is sufficiently sceptical of the short-term prospects for HDTV that it has proposed an interim solution.' This is enhanced definition television, which offers better picture quality than NTSC at a level somewhere between PAL and HDTV, with much cheaper transmission equipment than HDTV (a solution tried in the UK by the BSB satellite service).

HDTV PRODUCTS

Organizations including Thomson Group of France and the Science and Technical Research Laboratories of the Japanese Broadcasting Corporation, NHK, have developed products for broadcast TV production in HDTV. SOCS Research support 'near-HDTV' in their Quality Definition Television (QDTV) processing system. Philips, Pioneer and Sony make high-definition products for the non-broadcast market, some of which are used in projects described on page 75.

Philips and Sony demonstrated prototypes of their disc players at the International Broadcasting Convention in Brighton in 1988. Philips' HD-MAC machine was based on its VP-400 series of LaserDisc players, with four audio channels and up to 60 minutes of (non-interactive) full-colour video. Sony's HDL-2000 player supports the Japanese standard, with a choice of 15 minutes of non-interactive video, or eight minutes of IV on a specially-processed disc (incompatible with LaserDisc). The Sony player was released in the US in December 1989, for $2500 (£1500). In June 1990, Pioneer in Japan added a videodisc player to its CUBE Hi-Vision system, with 20 minutes of video per side, specifically for the Japanese museum and exhibition market.

HDTV IN USE

Some organizations are mounting the learning curve in preparation for the wider use of high-definition television systems. The BBC has built an HDTV outside broadcast unit which has covered the Football Association (FA) Cup Final, Wimbledon Lawn Tennis Final, ballet at Covent Garden, the Albert Hall proms and the 1990 World Cup. These programmes (some with surround-sound and multiple language commentaries) have been transmitted by satellite around Europe to demonstrate HDTV. The BBC has also used the Japanese format, notably in the drama series *The Ginger Tree*, made with NHK.

Despite this and other initiatives around the world, so long as the format war continues over HDTV standards, it is unlikely that there will be major spin-offs from the broadcasting industry into the interactive and visual presentation markets. For the foreseeable future, activity in the non-broadcast sector will only be of marginal interest to those promoting HDTV.

RECOMMENDATIONS FOR HDTV

Those who work directly with HDTV are vivid in their enthusiasm: John Galt, a former TV art director now working for Sony in Los Angeles, says succinctly that 'HDTV makes PAL look like shit'. The British *Screendigest* in September 1988 described HDTV as 'producing startlingly sharp pictures'.

However, enquiries made through the IMAGE museum network suggest that this enthusiasm is limited to converts. British producers of interactive media seem unlikely to become involved with HDTV in the near future, describing HDTV as 'a mess', 'a cul-de-sac' and 'a white elephant' – clearly, HDTV has not set the imaginations of these developers alight.

In view of the conflict in HDTV specifications, the very small amount of practical work now under development, and the negative response to the medium from established IV producers, anyone wishing to investigate HDTV, would be advised to:

1. Identify specifically the subject matter, applications and audience for which high-quality video is desirable or essential.
2. Appoint a member of staff or qualified consultant to monitor developments in the HDTV market.
3. Treat any involvement with HDTV as an activity running parallel to the main multimedia exhibition project, with its own project leader, resources, budgets and schedules.

Given the current state of HDTV technology and the inevitable lack of production expertise, the development of other displays should not be tied to what could easily become an expensive R&D exercise. Particularly, it is important to consider that a new range of digital video technology has emerged during the slow evolution of these various analogue systems, and the key to the AV markets of the future may lie in these truly radical innovations.

DARK RIDES & 'EXPERIENCES'

Dark rides and experiences are popular within the 'heritage industry' and attractions which address historical subjects with the emphasis on entertainment. But other institutions have incorporated experiences within gallery displays: the Black Country Museum in the West Midlands includes a model of a Victorian coal mine; the Imperial War Museum's displays are discussed on page 48.

Experiences are usually full-scale reconstructions which can range from a single set to a long panorama through which visitors walk at their own pace. Dark rides, by definition, seat visitors in some form of open carriage which moves slowly through an exhibition which may extend for some distance on either side of the conveyor. Audio CD and LaserDisc usually supply both sound and images.

The Disney parks in California, Florida and France feature some of the world's most lavish dark rides, where animatronics provide realistic moving figures of people, wildlife and even dinosaurs, holograms create spectral images in 3D, audio effects are cunning deployed from all angles and synthetic odours even suggest the smell of the primæval forest or the pirate's cannon.

A single dark ride or experience is often the centrepiece of a small theme park or heritage centre. Where the institution has little or nothing by way of a formal collection, modern reconstructions of sites, costumes and objects can be effectively displayed within this panorama. Heritage centres particularly tend to be richer in facts than artifacts, a dilemma which this sort of exhibition neatly addresses. The challenge is, of course, to strike a balance between entertainment and authenticity.

Media Projects International (MPI), whose other work is described elsewhere in this report, contributed AV production to a dark ride at Lake Garda, near Verona in northern Italy. MPI produced video and special effects to complement displays which also include animatronics and static figures in sets as complex as those of any Hollywood sound stage.

MPI's involvement, from a feasibility study through design and production, amounted to some two years' work, with a month spent installing the material on the site before its opening in July 1992. MPI Director Rosalie Vicars-Harris notes that management skills are critical in a project of this scale. 'There are lots of teams – many American – the client is in Italy, and the producers are all over the world. That calls for lots of coordination.'

This is a theme strongly endorsed by Colin Pyrah of Heritage Projects, a company which offers the full range of services from feasibility studies through research, design and production to installation, management and even marketing of attractions including dark rides and experiences. This represents a permanent staff of 45, including researchers, designers, project managers and workshops for scenery, costumes and props.

Heritage Projects grew out of Britain's first dark ride, at the Jorvik Centre – a celebrated attraction which simulates sights, sounds and smells of daily life in the Viking settlement at York. The company has since produced dark rides at the Scotch Whisky Heritage Centre in Edinburgh and The Oxford Story (where visitors sit in a 'scholar's desk' which moves slowly through a panorama of university life from the middle ages to the Edwardian period).

Their work also includes industrial heritage centres in Rhondda in Wales and the Black Country Museum, the White Cliffs Experience (where ancient Britons confront the Romans, and 20th century Dover suffers an air raid), a glimpse of Chaucer's England at *The Canterbury Tales*, a Georgian *Day at the Wells* in Royal Tunbridge Wells, the British Pavilion at the Columbus '92 exhibition in Genoa, and the largest temporary exhibition ever taken into the Soviet Union.

This gave the citizens of Kiev an insight into *British Life Today* with the reconstruction of a High Street (complete with Marks and Spencer, Boots and a Nat West bank) and a typical semi-detached family home. Other work abroad includes projects in Belgium, Ireland, Jamaica and Hong Kong.

Pyrah believes that the UK is ahead of Europe and the US in 'the design and management of well-organized heritage projects'. He emphasizes the importance of 'academic integrity' in designing exhibitions which are authentic as well as entertaining. 'The heritage industry in this country has suffered greatly by people just jumping onto the bandwagon.'

He also believes that it is vital 'not to design and build anything which cannot be maintained within the client's budget and staff resources'. Without proper maintenance, he says, these ambitious displays 'look tired very quickly'. Particularly, he recommends 'constant feedback', which Heritage Projects collects through exit interviews with visitors. 'Public expectations rise by the day', he says.

THE PRODUCTION PROCESS

The process of designing and producing interactive programmes is beyond the scope of this report. However, even where experienced programme-makers are entrusted to produce these displays, the client who understands the nature of the work and the technology can participate more fully in the creative process, and evaluate project management, proposals, budgets and schedules more accurately.

STAGES IN PRODUCTION

The stages in the production of any interactive multimedia presentation vary considerably according to its complexity, the platform on which it is developed and presented, and the experience of all those involved. The first IV and CD-I titles necessarily saw as much experimentation as achievement, and could take a year or more to produce. With more experience, better delivery systems, and a new range of hardware and software production tools, many titles can be produced within six or eight weeks, and ambitious projects in a matter of months.

Equally, with a new generation of hardware to capture audio, video and graphics as digital data, and tools to automate design and production, wholly computer-based displays can also now be created within weeks. Nonetheless, any production schedule should allow ample time for research before this work begins, and field testing before it is formally signed-off. Thus, any project might include:

- Preliminary analysis of aims, objectives and resources within the client organization (specifically, to determine how interactive multimedia systems can enhance conventional gallery displays and information services).
- Identification of potential hardware and software suppliers and services as appropriate.
- An invitation to tender, followed by proposals from potential suppliers, vetting of proposals and choice of suppliers.
- Detailed analysis to determine the identity, needs and abilities of the real target audience, the aims and objectives of the client, and the human and material resources required.
- Agreement of contract documents including budgets, production schedules, sign-off points and payment schedules.
- Research and design of programme structure and content, hardware specifications, including visual and operational conventions for both housings and programmes.
- Preparation of fundamental design documents including storyboards, scripts and flowcharts.
- Prototyping, typically with HyperCard or similar software which simulates the appearance of the larger programme, using static screens of text and graphics to illustrate its contents, and hypertext links to simulate its interactivity.
- Video production, photo shoots, archival research and creation of conventional artwork and computer graphics, as appropriate.
- Additional audio recordings and archival research as necessary.
- For multimedia platforms such as CD-I, CDTV or DVI technology, digital encoding of still pictures, graphics and audio, compression of moving footage and other AV material as necessary.
- Computer programming, often supplemented by authoring tools which automate parts of the production process.
- Integration of the various AV and computer components, initial testing and debugging, with changes and adjustments as necessary.
- Possibly, production of check discs for field testing, and further refinements as necessary.
- Pre-mastering, mastering and replication of LaserDiscs or CDs and/or computer disks.
- Ideally, implementation through staff training, promotion to the public and within professional circles, further field testing and formal or at least anecdotal evaluation.

Multimedia platforms may also include stages such as constraints analysis (to evaluate the design in view of the storage and processing capacities of the delivery system) and disc building and emulation (to simulate the performance of the CD on a computer for testing and amendment before the disc is pressed).

Optional activities include the design and production of housing for systems in public displays, documentation such as manuals and workbooks for staff or students, and packaging for software which is published commercially. These, naturally, run in parallel with the central production process, and often reflect visual conventions established there.

AUTHORING PRODUCTS

A whole body of software and hardware products now exists to facilitate the creation of interactive multimedia packages. These include equipment such as scanners and audio capture facilities, and software such as libraries of readymade code and engines to execute common computer routines. Complete authoring packages automate both the design and execution of the interactive presentation (for example, by giving authors the tools to create templates which they can fill with their own text and AV sequences, and simple instructions to create links in the interactive program, which the authoring package executes automatically).

These tools are, of course, used most efficiently by trained and experienced operators, but many allow specialists in other fields to apply their knowledge directly to the creative process. Some allow professional designers to create the package, and local subject experts to maintain and update it.

Multimedia, particularly with the added dimension of interactivity, demands a high order of skills, and considerable expertise in many areas beyond the subject matter itself. One person can indeed produce an interactive multimedia presentation single-handedly; however, the variety of skills and perceptions in a group usually informs the combination of instructional and visual design which is friendly, functional, attractive, engaging and satisfying.

SOURCE MEDIA

The choice of source media is critical to the quality of the finished multimedia production. Recordings from low-quality videotape will look pathetic, some generations later (that is, copied from one medium to another), by the time they reach LaserDisc. Conversely, computer graphics and platforms such as CD-I or DVI technology, with a choice of quality levels, can sometimes compensate for shortcomings in source media through creative production techniques. (Equally, Kodak's museum at George Eastman House in New York uses videodiscs to present photographs too delicate for constant display, and says that some pictures were 'enhanced in the video transfer process to improve image quality'.)

AV producers are divided over the relative merits of film and video, but the quality and versatility of professional tape systems, set against the cost of film, weights the balance in favour of video. Existing films or tapes can be used in a new interactive programme. However, this fundamentally linear material is not easily adapted to the demands of interactivity. It is often difficult to edit into the short segments typical of interactive displays, and linear material rarely allows for the many options and variations essential to interactivity. The production of complementary new material poses many continuity problems, especially when specific people or places are involved.

This is why IV projects usually end up shooting fresh video even if they began by re-editing existing material. An acceptable compromise can be seen in Longman Training's early programmes, which re-cut linear videos with new still frames, computer text and graphics, with voice-overs and fresh footage from *one* of the original performers to link the re-cut sequences.

'Re-purposing' works best with documentary or archival footage, an option which fortunately suits many museum and gallery displays. As well as continuity, there is the critical question of copyright, although this may be less of a problem for institutions like the British Library (which own both material and intellectual property, and the rights to exploit those collections) than for commercial programme-makers.

THE PRODUCTION TEAM

The variety of creative, administrative and technical activities described above could easily involve a large team of specialists, from project managers to computer programmers by way of designers of many kinds. Design alone may require expertise in instructional, interface, screen, video, graphic, text, print and housing design, and the firm guidance of a single creative director. Text screens, voice-overs and dramatic dialogue require such different use of words that one project may legitimately employ different writers to script or adapt the work of a content expert to the various demands of the presentation.

In practice, these demands are often met by people with the creative and technical skills to fill more than one role. Every project requires at least one subject expert but, clearly, there is much scope here for institutions such as museums and galleries to contribute specialists of other kinds to teams assembled by their suppliers.

There is a trend in corporate IV for clients to work closely with their suppliers to gain a foot on the learning curve. This at least qualifies them to assume project management, a role often taken by the client in other fields, but usually supplied by the programme-maker in this market, where a great deal of skill and experience is required to coordinate a project of any complexity. Organizations with in-house resources such as broadcast-quality video production facilities, or computer programmers with a background in computer-based training, may also take on specific technical roles.

Eventually, some companies have gone on to make programmes on their own; alternatively, some have commissioned programmes from external suppliers which they can subsequently maintain and update internally. Either way, now is the time to get on the learning curve, for there is too little formal training in this market even for professional designers, and experience is still the best teacher.

MAKING A CHOICE

Clearly, there are many options for the designers of exhibitions and displays in museums, galleries, marketing and information services. Few markets offer such a choice of radically different products for the same purpose. Interactive multimedia information systems begin with audio jukeboxes and stretch far into the realms of holograms, robotics and 'virtual reality', where three-dimensional effects seem to project the visitor into the very heart of the interactive display.

There are no simple rules for choosing multimedia hardware and software, suppliers and services – particularly where both technology and expertise are developing so quickly. However, a few general guidelines and observations may suggest some worthwhile considerations.

Although the distinction has been drawn between the domestic and professional markets, products designed for one may well be developed in the other. As even this brief summary indicates, digital media are developing rapidly in several directions.

This market is changing rapidly, but CD and digital video still have some way to go before they can compete directly with existing videodisc systems in terms of image quality and quantity. The analogue video which now supports broadcast television, videotape and disc will be with us for some time yet, particularly as it will take many years to convert the large international market for professional communications and home entertainment products to any new technical standard. However, this does not mean that digital media are not already a serious alternative to conventional systems.

This is an excellent time to enter the optical disc market, for there is now a real choice of products and enough information to make balanced decisions. At the same time, with a bewildering choice of incompatible platforms, there is likely to be some sharp competition between similar products, and a few commercial casualties. The closest comparison may be with the videotape market of the 1970s: after a long struggle, VHS eventually emerged as the most popular consumer product to the cost of those who had backed Betamax and other systems. However, a variety of formats survived on the professional market, where the wider range of applications can support a wider choice of products.

Fortunately, museums and galleries have more freedom of choice than, say, training establishments or education authorities. These institutions often need to accommodate existing systems, link terminals in local and wide area networks and integrate generic courseware made to standard specifications. Gallery displays, on the other hand, are often standalone, and need not communicate with other systems. Hardware and software may not be upgraded during the working life of the display. It can be most convenient simply to replace whole units as exhibits are redesigned, for a few years' service is likely to justify the hardware investment many times over.

Among the multimedia platforms, LaserDisc still has many strengths, particularly in the presentation of high-quality still and moving pictures. LaserDisc alone is a cost-effective alternative to conventional linear media such as tape/slide and videotape, and interactivity adds untold value to image bases and displays of all kinds. (Similarly, audio CD alone is an excellent, inexpensive and hard-wearing medium for music and audio of all kinds in interactive or static displays.) Both LaserDisc and IV are well supported by products, services and programme-makers with many years' experience.

Digital CD formats are catching up fast, but LaserDisc has one large advantage in that it effortlessly delivers what CD developers revealingly describe as full-colour, full-screen, full-motion video. This describes video as the rest of the world understands it: moving colour pictures which fill the screen unless they are specially processed to create effects like windows or artistically distorted images.

Digital formats are still most comfortable with video which either doesn't move (still frames) or fills only part of the screen (windows). Full-screen, full-motion video is something of a luxury as the compact disc formats now stand, for every minute of video eats up quantities of storage space and processing capacity which might be more economically deployed with still images, animation or text. Equally, many presentations have no call for moving footage, or can use windows to advantage to separate text from pictures, or display more than one image on the screen.

Systems which can compress large volumes of audio and visual material, such as DVI technology, can employ CD as a convenient and cost-effective medium for standalone terminals and networks. Equally, a large hard disk could store volumes of information that can be easily and economically updated. Thus, a large database of text and images could be stored centrally and displayed variously at terminals throughout the museum – for research by readers and browsing by visitors.

Recent enhancements to popular computer operating systems such as Microsoft's Windows 3.0 for the PC and Apple's System 7 for the Macintosh will dramatically extend the capabilities of the desktop computer. With products such as Multimedia Extensions to Windows or Apple QuickTime, users can now cut-and-paste still images, audio and video almost as easily as text or computer graphics, in applications software of all kinds. In the hands of good designers, this provides exciting new opportunities for museums and galleries to create their own multimedia displays entirely in-house.

The particular challenge of supplying images of high quality may be met with the introduction of Kodak's Photo CD. For little more than the cost of processing 35mm film, designers can assemble digital image bases which support displays up to four times sharper than those of high-definition television. As well as supporting a new generation of gallery displays, Photo CD also represents an unprecedented opportunity for marketing images from museum and gallery collections, with a single international format and players that may be found in home, schools and professional environments.

Most organizations, as soon as they decide to become involved with interactive media, immediately want to buy something. This is a common human reaction, and the cause of many unwise decisions, for a premature choice of delivery systems can seriously restrict the range of creative options subsequently available to programme designers. In multimedia, the final hardware choice should meet the requirements of the creative team, not dictate them. Most of the programme-makers interviewed here – and certainly all those whose working lives have taken them from books and slides to LaserDisc and CD – advise newcomers to concentrate on structure and content which can be adapted to new platforms as they emerge.

It is, at the same time, important to be aware of the capabilities and limitations of the systems currently available, and to keep abreast of market developments. As the project progresses and the specifications become more detailed, it will become increasingly important to base creative decisions on practical considerations such as cost or quality: knowing what the various media can achieve, and how different systems and devices actually work, may determine the feasibility of some creative ideas, and at the same time, identify those products most appropriate for the jobs at hand.

This may be disappointing advice for an organization which is keen to make a commitment, but the most important decision is that to produce interactive multimedia systems at all. The actual choice of hardware is secondary, and the sad truth for the technophiles in the project is that perfect transparency should leave most visitors quite unaware of the complexity of the technology behind the programmes which they have used and enjoyed.

SUMMARY OF MULTIMEDIA PLATFORMS

This chart summarizes characteristics of principal multimedia platforms discussed in this report. The acronyms CLV (Constant Linear Velocity) and CAV (Constant Angular Velocity) distinguish LaserDiscs which are essentially linear (CLV) from those which support interactivity (CAV). Obviously, the ultimate storage capacity of many formats depends greatly on the choice of encoding formats and quality levels at which AV material is recorded.

Format	Size	Type	Data storage	Media mix
LaserDisc	30cm	CAV	Digital/analogue	Linear text, graphics, audio, animation, ≈ 2 hours' video
LaserDisc	30cm	CLV	Digital/analogue	Interactive text, graphics, audio, animation, ≈ 1 hour's video
CD-DA	12cm	CLV	Digital	≈ 72 minutes' stereo audio
CD-ROM	12cm	CLV	Digital	650/675 MB data Text, graphics, audio, stills
CD-ROM XA	12cm	CLV	Digital	650MB data Text, graphics, audio, animation, stills, video*
CD-I	12cm	CLV	Digital	650MB data Text, graphics, audio, animation, stills, video*
CDTV	12cm	CLV	Digital	540MB data Text, graphics, audio, animation, stills, video*
DVI on CD-ROM	12cm	CLV	Digital	Text, graphics, audio, animation, stills, video
Photo CD	12cm	CLV	Digital	100 still photos (audio and video upgrades to follow)

* At time of writing (April 1992), video on these platforms is restricted to partial-screen, partial motion displays, but all promise full-screen, full-motion video in future upgrades.

BANK OF ENGLAND MUSEUM, LONDON

The Bank of England Museum, in the City of London, opened in November 1988. The museum occupies a corner of the famous building on Threadneedle Street, but has its separate entrance in Bartholomew Lane. This, almost metaphorically, is one of the museum's unique challenges: to present objects and places within the same building, to which public access is utterly prohibited.

The new museum replaces a smaller collection which could only be viewed by appointment and had a long waiting list. The Bank could not then satisfy public curiosity about its principal assets – its holding of the nation's gold reserves, and the rich interiors of its principal rooms. Neither could traditional displays adequately explain and represent the nature of modern banking, or the Bank of England's role in the nation's economy.

Thus, the new museum complements four suites of historical documents and artifacts with a mock-up of a modern dealing desk and a room of interactive terminals presenting *The Bank Today*. The attractor loop displays the Bank's famous doors; when this screen is touched, the doors swing open to display a menu of small labelled photographs which represent the nine 'rooms' of information in the interactive presentation.

Some of these represent actual locations within the Bank or its associated facilities – the Main Entrance and Court Room, for example, the Dealing Room, Printing Works and, of course, Gold Vaults. Others are metaphorical, and explain concepts such as Money and Monetary Policy. 'The concept of banking is very cerebral, very esoteric', says John Keyworth, of the Bank's Museum and Historical Research Section, 'How else can you present a subject such as Banking Supervision?' The Supervision 'room' includes footage of the Bank of England's representatives visiting other banks, and explains what its employees actually *do* to keep the UK's financial institutions on an even keel.

The final room is a trivia quiz. Ravi Singh of Chrysalis Interactive – who produced the programme when they were called Blackrod Interactive – says this was designed for children but has proved popular with City business types, who appear to enjoy anything with a competitive edge. (This phenomenon has been observed in training, where financial institutions particularly have found that staff disdain the confidentiality implicit in self-paced, technology-based learning, and eagerly compare scores. It is also observable where school groups gather around interactive displays such as *The Living Arctic*, and can be positively exploited with cooperative rather than competitive games.)

The Bank of England Museum's various presentations make extensive use of what Singh calls 'The question that informs: "Did you know that ...?" It's a more interesting way to learn than being deluged with facts.' John Keyworth personally has reservations about the amount of information which visitors actually absorb from watching the multimedia programmes, at least by comparison to reading the substantial and informative labels on the conventional displays. Nonetheless, he says that the interactive displays convey ideas and information which could be 'quite boring' in other presentations.

In its first three years, the museum attracted 310,000 visitors. (When the museum opened, the Bank's representative observed that it was largely a PR gesture as 'There is a steady demand for our products without advertising'.)

Keyworth says that the interactive displays are second in popularity only to those of solid gold bars. He reports no more than the 'odd hiccough' in the hardware. (The delivery system is based on the Sony VIEW, which combines the LaserDisc player and computer in one box.)

The systems are housed in a small, busy room which provides an implicit comment on the changing face of finance in its very contrast to the traditional displays elsewhere. The mock-up of the dealing desk uses a telephone handset and wide-screen display of a dealer at work. The touchscreen terminals are complemented by larger passive monitors, so visitors on busy days can at least watch the presentation which someone else controls. However, within the constraints of the building, these ceiling-mounted monitors are within reach of tall adults, and it was interesting to watch journalists at the Museum's press launch futilely tapping the TOUCH HERE instructions on the monitors above, as the terminals stood unattended below.

Keyworth notes that the museum chose another form of optical disc, audio CD, to introduce unobtrusive music, and add commentary to a display on the storming of the Bank during the Gordon riots of 1780 (visitors simply press a button to hear a 90-second narration). Interactive displays have clearly enhanced a collection which was intrinsically interesting, and is handsomely displayed.

BIRMINGHAM MUSEUM, GALLERY 33

Gallery 33 is a discovery room which opened in the Birmingham City Museum and Art Gallery in 1989. *Collectors in the South Pacific* is an interactive LaserDisc programme which combines information on the history of a collection with insights into its interpretation, and provides a database which visitors can explore on their own.

The interactive material presents contrasting interpretations of history and culture by examining four artifacts from the museum's Melanesian and Polynesian collection. One aim of the discovery programme is to alert visitors to the idea that the objects gain their value by the meanings we impose upon them, and to make them think fundamentally about the nature of collecting and the display of artifacts in museums.

The four 'guides' to the South Pacific collection represent the different perspectives of a missionary, a collector, a tourist and a museum curator. All four are 'real people' and the two historical figures have local connections: the missionary was Ida Wench, a native of Leamington Spa, and the collector was the Birmingham-born Arthur Wilkins, who emigrated to Australia but bequeathed his cabinet of curiosities to his native city. Of the two modern characters, the tourist is American, and the curator comes from the Solomon Islands' National Museum.

The system uses a touchscreen display, with still and moving pictures, archival footage, voice-overs and captions. Images of the four artifacts appear in windows in the centre of the screen, with photos of the four guides in the corners: visitors touch the screen to select an image and a guide, to hear different information and perspectives. Users can select as little or as much as they like, and consider the evidence in any order.

Thus, the curator will explain the cultural significance of native jewellery, and warn that these objects cannot legally be taken from the islands, while the tourist knows how much they cost in American dollars. The collector relates how he originally acquired and displayed the objects, and the curator tells how some foreign museums are now returning artifacts to the islands' own collections.

The tourist says that she expected to be greeted by colourful natives in outriggers, but found the waters full of Japanese factory ships, observing with some interest that they can catch as much tuna in a day as local fisherman once took in a month. There is no overt commentary on any of these facts or opinions – users are left to draw their own conclusions.

The presentation also asks visitors to consider how museum collections should be catalogued and displayed, by category or context. These ideas are supported by an image bank on the LaserDisc of 117 further objects, which can be explored in various ways. As well as the familiar curatorial terms such as provenance or collector, users can pursue themes with more general appeal. This allows them to find, for example, all the objects which represent animals, or depict images of the sea – perspectives which many visitors (particularly children) find more attractive and interesting than scholarly classifications such as medium or period.

The idea for the programme came from Jane Peirson Jones, Keeper of Ethnography and Anthropology at Birmingham City Museum and Art Gallery. The package was produced by New Media, with sponsorship from the English Tourist Board, the Gulbenkian Foundation, Central Television and the Friends of Birmingham Museum.

Within the gallery, the combination of an active terminal with a passive monitor above allows passersby and larger groups to watch the screen while individuals or small groups interact with the system. In the spring of 1992, the museum was in the course of assessing reactions to displays including *Collectors in the South Pacific*; the results of this survey should be published in due course.

New Media's Nicholas Lewis speaks of the interactive programme as 'creating a mood' rather than cataloguing a collection. It approaches objects not with the values of curators and scholars, but from the perspective of visitors – and, particularly, children – who are more likely to be interested in looking at faces or patterns than considering sources or probable dates of origin. From the observable response to the system in Birmingham, it is clear that museum visitors can discover much on their own, when they are given the means to explore.

Writing in *Museum Development* (June 1990), Jane Peirson Jones observed: 'Interactive video [is] an excellent medium in which to present different perspectives on a particular subject .... it lends itself to exhibits which aim to be open-ended, which raise questions and which actively involve the visitor'.

BRITISH GOLF MUSEUM, ST. ANDREWS

'Where you have point-of-information, you also have the potential for publishing.' The British Golf Museum and I Contact have produced the world's first installation of CD-I in a museum, and adapted this for publication on disc.

Great British Golf is a commercial CD-I title based on material developed for a new museum at St. Andrews, Scotland, the birthplace of golf. It incorporates popular features of the displays in a three-part package which combines a history of the game with a trivia quiz and a simulated game on the Old Course with historic balls, clubs and players.

BRITISH GOLF MUSEUM

The British Golf Museum uses eight LaserDisc and three CD-I systems to support an AV theatre and present interactive programmes produced by I Contact from scripts written by the museum's Director, Peter Lewis. The museum is arranged on a winding path which encourages visitors to view the displays in the order determined by their designers. The interactive AV package is controlled so that each terminal displays only material relevant to the adjacent exhibits.

The museum also installed six slide projectors, specified by the exhibition designer 'against the advice of Philips and I Contact', says Lewis, who observes 'We have experienced more problems and down time with the projectors than with all of our optical disc players put together'. CD-I now supplies the stills in place of slide projectors.

The limited video facilities of the first-generation CD-I players are met by the extensive use of windows to display boxes of moving footage within the larger screen, but the package is designed to accommodate an upgrade to full-screen, full-motion video in the future. The video clips combine modern shots of St. Andrews with archival footage, which necessarily includes a good deal of grainy, and even fragmentary, black-and-white film and early colour stock. Much of this material positively benefits from display in windows, which minimizes the contrast between clips of radically different visual quality. In one novelty item, the individual pages of a flicker book were shot as stills, and run at 14 frames per second to simulate the effect of the original illusion.

In its different incarnations, the package has been developed for use with a trackerball, touchscreen and the CD-I player's own thumbstick controller (a hand-held remote control device incorporating a miniature joystick). I Contact observed that, whatever their visual-spatial and athletic skills on the golf course, many visitors to the museum found the trackerball challenging. However, Peter Lewis told the second annual conference of CD-I developers in London in 1991: 'When confronted with the first touchscreen, visitors are often shy and tentative, but by the time they get to the fourth one, wild horses can't drag them away'.

The interactive terminals alternate in their use of audio and silent presentations, so that ceiling-mounted speakers are strategically placed at least two metres apart. This usually guarantees that visitors around any one terminal only hear the sound appropriate to that display, but teething problems with the cabling (which was not shielded) have occasioned a few surprises when individual tracks reached the wrong speakers.

Like several other developers interviewed here, I Contact endorses close working relationships between architects, exhibition designers and external suppliers of display material. The installation of the multimedia systems involved some scenes not dissimilar to those described at the National Museum of Racing in Saratoga Springs (*q.v.*) – again, because of simple disparities in the planning of the various facilities in the museum. Separate control rooms (of disproportionate size) were provided for the linear and interactive hardware which feeds the terminals, and neither was within easy access of the displays themselves. Surrounded by high technology, the I Contact team effected the last stage of testing by shouting loudly across the museum to explain to one another what was happening in the control room and on the screens.

Mark Bonsor of I Contact strongly recommends 'taking lots of spare kit for the installation', particularly to remote sites. (This museum is supported with a service contract from a supplier in Newcastle.) He also endorses field testing of every aspect of the installation: at the Golf Museum, the original plan was simply to leave the systems playing 24 hours a day, as this no longer taxes the hardware or software unduly. However, heat and noise from the players tended to set the museum's sensitive security alarms ringing in the dead of night, so there is now a daily start-up and shut-down routine.

Museum Director Peter Lewis worked closely with I Contact in the development of the packages, as their content expert (and, I Contact report, 'the world's worst golfer'). From his observations of interactive systems, he believes that 'Familiarity breeds contentedness. The visitor should not have to go deeper than a main menu and one sub-menu to access a programme, and they must be able to stop at any time and return to the main menu either directly or via one sub-menu. This should be very well sign-posted on the screen'.

In writing his own scripts, he also quickly discovered that 'The time spent watching an AV is deceptive. When I went to one museum to look at their use of touchscreens, I thought that the AVs were lasting about three minutes. When I started to time them, they were actually lasting between 70 and 90 seconds'. He believes that, 'On average, museum visitors only spend something like five seconds reading a text panel' and aimed to keep the AV clips in his own programmes 'to about 60 seconds – my simple guideline was that no script could fill more than a single page of double-spaced paper. I eventually discovered that professional actors read more slowly than museum directors, so most of the AVs are actually about 70 seconds long.'

I Contact went on to win contracts from other sports museums in Europe and America, and to develop other entertainment titles on CD-I.

GREAT BRITISH GOLF

I Contact were trading as APS Interactive when this package was developed, in an organization which comprised a below-the-line advertising agency and a multimedia production company. This provided in-house studios with many resources which other companies must hire, including traditional cell animation as well as a suite of Macintosh computers using Adobe PhotoShop for stills, MacroMind Director for animation, and HyperCard for prototyping. This in part facilitated the development of different versions of the British golfing disc.

The original museum material took well over a year to develop since, as one of the very first CD-I projects, much research accompanied the development. The consumer disc was also one of the first, but was produced in nine months, with the work of 28 people overall (including six software engineers, two credits each for audio and text, three managers for different parts of the production facility, as well as designers, director and producer).

The consumer disc employs an attractive collage of Victorian images as a background to a scrolling menu which is actually four times the width of the screen display. Users pan back and forth across the main menu to select the introduction or exit, the history, trivia quiz or game.

Hotspots such as these main menus change colour as the cursor passes over them, to indicate where there is more information or an opportunity for interaction. Selecting any topic on the main menu produces a pop-up sub-menu where users can scroll up and down a list of contents, to highlight and select the presentation of their choice. Clicking as the cursor passes over the drawing of any one golf club, for example, produces a small text display with more information about it.

The consumer disc includes 98 AV clips, a database of facts and statistics, a trivia quiz and a computer game. The trivia quiz contains over 500 questions, which players answer in rounds in 18 'holes' per game. They play against a clock on the screen, to score par, birdie or eagle, the quicker they provide the right answer. The CD-I system can hold scores, so players can monitor their personal bests by comparison to par.

A sequence on 'what made great players great' is virtually a tutorial: I Contact Marketing Director Steve Baker (who has recently taken up golf) says that watching archival footage of Bobby Jones in action is a model for any modern golfer.

In the simulated round on the Old Course as it was, players can choose historical figures who actually played at St. Andrews – Mary, Queen of Scots, the club's first Captain or first pro ('Old Tom' Morris), each represented by an animated cartoon-style figure. There is a choice of three balls, each with its own liabilities: the feather (which can explode on contact), the gutty ('not as ærodynamic as modern balls') and the haskell (prone to carry air pockets). The choice of clubs runs to the brassie, cleek, mashie and niblick, and the hazards include grazing sheep.

This local colour is exacerbated as 'a caustic Celtic caddie' comments on each player's performance. (The voice for this character was supplied by David Anderson of Philips, a detail which the LaserDisc fraternity should relish.) The caddie is also an animated cartoon character, with a bushy moustache and omnipresent fag-end which effectively obviate the need to lip-synch his comments, as his mouth is hardly visible.

I Contact are as curious as anyone to see how all this will be received in Japan, where enthusiasms for both golf and new technology predicate a healthy market. The Japanese version is being developed bilingually, as many golfing terms are taken from English, and many Japanese golfers are English-speakers who were drawn to the sport by their international contacts. The I Contact team have found that the hardest part of translation is pacing, whether words are written or spoken. The caddie's voice is supplied by a popular comic actor who can speak Japanese with a Scottish accent; however, there is no precedent in Japanese society for the caddie's cheeky causticity, not the least because Japanese caddies tend to be extremely polite women.

BRITISH MUSEUM, LONDON

The Making of England explored themes in Anglo-Saxon art and culture in a temporary exhibition jointly presented by the British Museum and British Library in 1991/92. It was also the impetus for a unique project conceived specifically to develop an interactive resource package for schools from material originally offered to museum visitors.

The information system was lodged in a wide corridor which led from the exit of the exhibition to the rest of the museum, thus providing not a gallery guide but additional interpretative material. Visitors were advised that it represented early stages of development which, if successful, could be produced as a CD-ROM for schools, and were asked to complete a short questionnaire to help with the preliminary evaluation.

Four presentations – *Digging Up London, Monks and Nuns, Making Books* and *Marvellous Objects* – experimented with a variety of design features and interactive strategies. *Marvellous Objects*, for example, encouraged visitors to examine close-ups or compare objects of similar design or period. *Making Books* comprised 10 uniform screens of text and pictures which visitors could read sequentially or select randomly from a text menu. *Digging Up London* demonstrated the use of primary sources – to locate a site from descriptions in an ancient charter, for example, or reconstruct a building from a detail of the Bayeux tapestry. *Monks and Nuns* used multiple choice questions to introduce fuller explanations of daily life and religious practice.

The various segments used icons illustrated less metaphorically than historically – a leaping stag for START, and a hare for GO BACK – with ? for TOPICS, and a hand alongside the hare for CONTINUE. Neither the images nor the words employed conventions familiar from other interactive systems or computer software (?, for example, often represents the help function). Thus, it seemed that many visitors did not perceive that the stag or START actually represented the main menu from which the four separate presentations began, and could not easily find their way from one section to another.

(At the same time, when a small bug in the software occasionally revealed an arrow-shaped cursor on the screen, users responded promptly to that familiar symbol. The cursor was a throw-back to the mouse with which the system was programmed, and had no function in the touchscreen version, but the common association of an arrow with action prompted users who knew all about the buttons and hotspots to touch the arrow instead.)

Nonetheless, it was observable here – as with many other systems described in this report – that visitors who were clearly not computer-literate were sufficiently confident to experiment with the system and everything they found there. The use of buttons, hotspots and menus posed challenges to some, which most seemed willing to tackle – often, cooperatively, as friends would work together or small groups would form around the screen. Some only knew what they wanted to see because they had glimpsed tantalizing pictures during someone else's time at the screen. Clearly, this is a select sample – visitors inquiring enough even to find their way to the screen, and persistent enough to wait their turn and seek specific information – but where there was an incentive, some seemed willing to grapple with virtually any obstacle.

Making Books, in particular, foxed several with its pop-up scrolling menu – a device common enough in Macintosh and Windows software, but without precedent in everyday life. To use this menu, or even to get rid of it, visitors had to grasp the principle of running a finger up and down the vertical bar at the edge of the menu to see all the topics, and touching one of two buttons at the foot of the scroll to select a topic or CANCEL. Some only achieved this by touching virtually everything on the screen – but that, too, is a testament of sorts to the resilience of the hardware and software, and the insouciance with which many people now treat anything that looks like a consumer product.

The next phase of the project will develop the package as a resource for schools. This is possible, in part, because the project was conceived with both objectives; it is certainly facilitated by the fact that the British Museum and British Library own so much of the source material and intellectual property which such an initiative demands. (Some of the other developers interviewed here came to this idea too late, and found that they could not extract or develop educational material from the structure and contents of their public information systems.)

The project began with a spot of matchmaking by the Computers in Teaching Initiative (CTI) Centre for Literature and Linguistic Studies of the University of Oxford . The core team emerged from links between Marilyn Deegan of the CTI, Rowena Loverance of the British Museum Education Service,

Michelle Brown of the British Library's Department of Manuscripts, and two suppliers – Research Machines (RM) and Office Workstations Ltd. (OWL). RM supply a niche market with Nimbus computers, which bridge the gap between the ubiquitous PC and systems developed for British schools in the early days of desktop computing and IV. RM also distribute CD-ROM drives and OWL's Guide hypertext software – a product range which effectively specified the platform for this project.

Loverance recognized that 'We were faced with an exhibition which we knew would be highly academic, with a wide range of audiences, and we were actively seeking new ways of presenting material'. In particular, the Education Service was keen to address seven-year-olds, who study the Anglo-Saxons as part of the Primary History Curriculum. Multimedia was an attractive complement to initiatives which included an opera project with six schools in Suffolk, and a handling collection of artifacts and replicas which school parties can investigate alongside the interactive system in the Education Service's own rooms.

Loverance found that *Monks and Nuns* proved particularly popular, but has also observed that children quickly make the connection between images on the screen and objects which they have seen in the exhibition or handling collection. She felt that the interactive system has been helpful with concepts that are 'jolly hard to get over to children, such as how the Anglo-Saxons changed over time, or the use of primary sources'.

Chris Jennings of Research Machines acted, he says, as 'a sort of *animateur*', working with the three subject experts to help them grasp what an interactive system could do, and how they could employ multimedia effectively. 'The authors – that is, the content providers – need to know what is possible', he says, 'They couldn't give me material until I had given them the structure for it'.

Jennings began 'by scanning vast amounts of images without really knowing what we were going to do with them'. This gave him the resources to create prototypes which, in turn, gave the authors a concrete model which they could then enhance and refine. It was easier for the subject experts to provide information when they could perceive the structure in which it would be presented. 'I value the expertise of museum staff in two ways', Jennings says, 'in that they are subject experts, but they are also educationalists'.

Loverance says that, from extensive experience in developing materials for print and illustrated lectures, she found it 'fascinating to see what you can do in this medium'. On the technical side, she noted that many images produce quite different effects in print, slides and digitized screen displays, and every image (whatever its quality at source) must be individually evaluated on screen.

Chris Jennings produced the package, using a scanner to capture the still images and Guide to create the interactive program. 'It's a lot of work, but it's possible,' he says. Of all the developments in the multimedia market over the last decade, he agrees that 'it's really the quality of the graphics that have changed everything – you can comfortably read text and look at pictures on the computer screen now the way you never could before. With IV, you can't really do anything but go out and shoot professional video. This project, on the other hand, required no special photography – you can go to existing resources and use what they have.'

The emergence of CD and digital audio, he feels, also provides excellent opportunities for cost-effective educational packages. 'Interactive audio is wonderful, and it's not that hard to do.'

Jennings argues that ordinary PCs, with the option of a CD-ROM drive, still represent the largest established market for multimedia publishing. He notes that the most impressive applications of the Macintosh – in the Micro Gallery or *Sculpture Interactive*, for example – require fast, powerful machines with plenty of storage. Even the Multimedia PC specification calls for equipment which 'very few people have on their desk'. He recommends that developers concentrate on packages which make minimal demands on the hardware, and view even digital audio as an optional feature (at least until more PCs have audio facilities), if they want to address large audiences, especially in schools.

The project began in May 1991, and opened with the exhibition in November – a schedule which obliged the team to concentrate on four topics. The expanded version should embrace other themes from the exhibition and, thus, other subjects in the National Curriculum. *Kings and Kingdoms, Imaginary Animals, Myths and Legends* and *Places* could offer resources for English language and literature, Geography and Art and Design, at the least. Loverance has already identified the Romans (an important topic in the Secondary Core Curriculum) as a potential package for future development; equally, given the difficulty of siting interactive systems in busy exhibition halls, she notes that it may be easier to design a system for the museum's new oriental galleries than to find space elsewhere.

For schools, Jennings says, the basic presentation package must be enhanced with 'more topics, more material, more images and more primary source material. In a museum, if people stay in front of the systems perhaps five minutes, a crowd will form. In a school, it's got to be a lot more substantial because it will be used over and over again by the same group of people.'

CAMBRIDGE INTERACTIVE

The Nagas was developed by the Cambridge Experimental Videodisc Project, who worked with several British museums and scholars to assemble artifacts, photographs, maps, books, diaries, articles and other documents which describe the life and culture of the Naga tribes of Assam. This diverse archive, captured for storage on LaserDisc and controlled by a proprietary database, became one of the first multimedia databases. A brief description of *The Nagas* may help those unfamiliar with such products to appreciate how a simple data retrieval system can transform the use of audio-visual material.

The Cambridge Database System 2000 (CDS 2000-V) was originally developed within the University of Cambridge to create and maintain large data sets, and passed through five generations in 20 years, evolving alongside smaller, faster and more powerful machines to serve a growing body of computer-literate users.

The database was adapted for use with LaserDisc and other AV media by Cambridge Multimedia, whose founders first worked within the University, and maintain close contacts there. Cambridge Multimedia have since developed IV and multimedia programmes for training, education and public information, and have designed other image bases and databases for clients in museums, manufacturing, medicine and the financial sector.

CDS 2000-V combines free text and structured Boolean techniques (that is, selective searching through logical operators such as 'and', 'or' and combinations thereof). With *The Nagas*, for example, the team who prepared the database used keywords to describe salient features of AV material on the disc. A search for, say, references to 'women' and 'water' reveals the total number of references to 'women', then to 'water', and then to 'women' and 'water' in the same entry. Users can scan the text references and mark those of interest. At any point, the user can also display associated material from the disc. The computer text can be displayed on the same screen as the video, but a two-screen delivery system more effectively presents the full video image (still or moving) alongside the textual reference which accompanies it.

Users can change their terms of reference at any time, by adding new search words or deleting old ones. The 'semi-intelligent' system itself can deduce other possible keywords from the frequency of their repetition in the entries which the user has marked: 'women' and 'water', for example, might lead to 'bamboo', since Naga women often carry water in bamboo tubes. To make the results of multi-word searches more manageable, users can enter the most important word first, the least important last, and the system itself will arrange the selected records to meet these criteria. 'Women', 'water' and 'bamboo', for example, could be arranged to start with references to 'women' *and* 'water' *and* 'bamboo', and finish with references to 'bamboo' only. The user who then became interested in the use of bamboo could of course readjust the search criteria accordingly.

Users can save their findings from the text database as independent computer files, to resume work at a later date, print a catalogue or export the file to a word processor or other applications software. They can also use their findings to create separate programs for an IV presentation from the disc.

CDS 2000-V runs on an IBM XT/AT-compatible PC with a VideoLogic interface card. It can search 20,000 records in a 30MB database in less than 10 seconds, using about 300K of RAM and 500K on the hard disk – which makes it, as computer software, both fast and small. There is no limit to the number of records, fields per record, or words per field, providing that no record exceeds 64,000 characters (*i.e.* about 13 pages the size of this one).

Of course, the value of any database depends on the quality of its records and, thus, on the people who design and maintain it. Critical elements such as the choice of fields, the accuracy of factual and descriptive entries, and the choice of standard terms and spellings, rest entirely with human expertise.

Twenty years' observation of librarians, curators, archivists, researchers, scholars and students has led to modifications of the CDS 2000-V package. It is, for example, structured so that records cannot be altered by ordinary readers, but authorized users such as curators can easily add, delete and change the text as necessary – for example, to include new references to visual details which were not described in the original records.

CDS 2000-V is used at Cambridge on mainframe computers in the Scott Polar Research Institute Library and the Sedgwick Museum of Geology, for a video library catalogue in the Department of Social Anthropology, and with *The Nagas* at the Museum of Archæology and Anthropology. Other users include the North of England Open Air Museum.

CARTWRIGHT HALL, BRADFORD

'It's simple, it's lively and it works.' This was one reaction to the interactive information system which made its debut in Cartwright Hall in March 1990. The single kiosk stands in the entrance of the Hall, continuously cycling a 60-second attractor loop of colourful pictures which invites visitors to touch the screen and find out more.

The unit is designed and sited to make the imposing entrance of the museum more attractive to first-time visitors. Staff had observed that many people would actually reach the very door, only to fall back at the Edwardian opulence of the entrance and ground-floor sculpture court. The bright, modern interactive unit was designed by Media Projects International to act as a user-friendly bridge between the familiar world of television and the less familiar environment of museums and galleries.

The 20" monitor is mounted at a height accessible to older children, and not uncomfortable to adults. A single message – PRESS ANY BUTTON TO START – appears both on the screen and on a panel mounted immediately beneath. Visitors use eight bright yellow buttons to select numbered options from the screen, and darker directional arrows to page through entries which fill more than one screen. These are clearly marked 'Page x of y' at the foot of each screen, so users know when to expect more information. The programme was made in English, but use of numerals and pictograms on the keys allows for conversion to a multilingual presentation without alteration to the housing.

The system contains information about Cartwright Hall and its activities, and about other museums, galleries and events in the area – including an impressive map studded with places of interest – with instructions for travel to each site by car or public transport.

The system was to be updated every three months, to add news of forthcoming events, delete old notices, and modify data such as opening times or prices. In fact, as the museum plans major events fully a year in advance, the system has only needed updating twice-yearly. A bright red moving text display is mounted above the interactive unit, and staff at the Hall can program this at any time to add news of special events and daily attractions. A large noticeboard and leaflet racks stand immediately beside the system.

The presentation combines still photos from LaserDisc with text from a plug-in EPROM. This erasable, programmable device obviates the need for a computer: data and control programs are stored on the cartridge and interpreted by a microprocessor, which also controls the videodisc player.

To enhance the appeal of screens comprised mainly of computer text, considerable attention was paid to their design – particularly, the amount of text displayed. All text is laid over an attractive background recorded as a still frame on the videodisc. This combination of media makes best use of both – the visual quality of the disc, and the convenience of the EPROM.

Nearly two years after its installation, Jane Whittaker, Arts Marketing Officer at Cartwright Hall, reported that the information system has proved extremely popular with visitors of all kinds, and has become an integral part of the museum's new image. Attendances have increased 19% – and while the interactive system can't take all the credit, it does observably draw younger visitors over the threshold. 'Once they've pushed a few buttons and looked through the door', she says, 'they stay and they enjoy themselves. The moving panel has been brilliant for day-to-day events and concerts – and to tell them that admission is free. It gets them through the door!'

One early 'hiccough' was caused not by high tech hardware, but a loose wire. Crowds of excited children pressing buttons created enough vibrations to work the wire free, and bring the whole system to a standstill. The wire was secured, and the system has worked well since.

'Children get the most out of it', Whittaker says, 'because they're the best at pushing buttons. Older people walk past watching it and waiting for it to tell *them* what to do.' Adult visitors, she says, will find what they want from the system, and then turn to an attendant to confirm what they have just learned.

(The observations of staff in museums, galleries and attractions of all kinds confirm both that children are much more adventurous and competent in their use of interactive systems than adults, and that adult visitors particularly often want to confirm what they have learned from a machine by talking to a person, even one who knows less than the interactive database contains.)

Cartwright Hall has also added piped music to the galleries, and a laser-based counter to record attendances. 'Entertainment and visitor satisfaction are the important trends', says Jane Whittaker.

THE DESIGN MUSEUM, LONDON

The Design Museum in Butlers Wharf, on the south bank of the Thames opposite the Tower of London, has two interactive displays – one using an IBM-based interactive LaserDisc system, the other combining digitized still pictures with text on an Apple Macintosh. One of London's newest museums demonstrates that media choice is ultimately a question of fitness for purpose.

The Design Museum covers the history, theory, practice and future of industrial design in three main sections: the Boilerhouse presents changing exhibitions, the Review provides a shop window for new ideas and products, and the Study Collection displays some 400 objects illustrating the development of mass-produced consumer design.

Both interactive presentations are arrayed, with several terminals for each, in an open area adjacent to the Study Collection. A third AV system there presents a continuous loop of interviews with contemporary designers, and additional systems with short linear programmes are used elsewhere in the museum. Special exhibitions sometimes also employ interactive information terminals.

DESIGN A TOOTHBRUSH

One of the most popular exhibits at the museum attracts at least one visitor in five to a computer-assisted design (CAD) package which allows users to *Design a Toothbrush*, using interactive LaserDisc. Users must specify the materials, shape, dimensions and colour of the toothbrush to satisfy the conflicting demands of æsthetics, ergonomics and manufacturing.

The project has sign-off points, at which those who do not meet the specifications are asked to 'reconsider' their decisions. There is a budget, and a 'money clock' on the screen will terminate the unsuccessful designer's attempts if funds run out before the job is finished.

The project was a joint venture between the museum, the Royal College of Art and Sony UK. Sony's sponsorship took the form of a donation of VIEW delivery systems, extra monitors, a camera and a screen for the museum's lecture theatre and workshops. The RCA's computing department prepared the control program, and its students shot the video sequences. The whole package cost about £50,000 to produce in 1989.

The CAD system is controlled entirely through a touchscreen – users simply point to options presented there. The museum reports that there have been no problems whatsoever, either technical or conceptual, with the touchscreens in this application, or the mouse in the database described on the next page, despite long hours of constant use by adults and children alike.

Eight stations are dedicated to toothbrush design, with the IV units mounted on neat tables, and headphones provided for each user to reduce the ambient noise. In the summer of 1990, a year after the Museum opened, Systems Manager Hariet Phillips said that the programme was so popular that 'you'd be lucky to get a seat on most days'.

The Design Museum is unusual in *offering* seats: most information units, in museums and shops, expect users to stand in front of the screen. Seating encourages longer individual use of the system – this exercise seems to take most visitors at least 15 or 20 minutes, and some will play more than once, to try different strategies. This, of course, reduces throughput – many point-of-sale terminals expect to engage the user for no more than three minutes at a time – but it also gives visitors a more rewarding and memorable experience.

In practice, the game is best played just once. The money clock is so generous that it is actually difficult to run over budget, although the first-time player is unlikely to notice this, and should be gratified to achieve a healthy 'profit' at the end of the job. Design decisions are selected from three options, the aptness of which presumes that the designer already knows a good deal about toothbrushes. In the choice of materials and colours, for example, only one choice satisfies the criteria, while in the design of the head and neck, more than one combination is acceptable. However, the user is not to know this unless playing the game a second time or comparing notes with another player.

Users actually learn little about the mysteries of toothbrush design *per se*, but do get a good overview of the many issues and relationships which inform industrial design. They are obliged to view a good deal of supplementary material about the nature of design and the manufacturing process every time they play the game, while the money clock continues to run up the score – which is hardly a reflection of how a real design budget is spent. But these features are really only distracting the second time round, and a single round of the game is exciting and entertaining. The video sequences, of a trendy young designer at work, are already worthy of a time capsule.

STUDY COLLECTION DATABASE

The Design Museum's second interactive display provides a resource base for objects in the Study Collection. Many of these are on loan, so it is anticipated that fully a quarter of the collection will change every year.

Museum visitors, students, academics and professional designers approach the collection in different ways – some simply to browse, some to conduct a systematic analysis of one designer, manufacturer or movement, product or material. A computerized hypertext system, which allows users to make free associations through all parts of the database, offered a radical alternative to traditional caption cards and catalogues.

This facility had to combine accurate representations of the objects with fairly large volumes of text – a vexed point in screen design, as several of those interviewed here will attest. It was important to produce a system which was responsive, inexpensive and easy to update; however, truly photographic images were not required, as visitors would be encouraged to study the objects themselves in adjacent display cabinets, and use the database as a complementary resource.

The database displays digitized still pictures on a standard Apple Macintosh SE with a monochrome screen: a typical screen is illustrated here. The system is silent, which suits its purpose and allows the siting of several units in close proximity. Six menus introduce designers, manufacturers, movements, products, and indexes by country and of all the entries in the system. A mouse on the desktop is used for browsing and choosing.

Hypertext links allow users to move between entries – from information about designers or manufacturers to their products, for example, or from a product to its designer or manufacturer. They can also explore associations between people, ideas and objects by selecting keywords on the screens to follow selected themes through the collection.

The specification was prepared by museum staff and Cognitive Applications. The museum knew roughly what they wanted the system to do, and whom they expected to use it, Cognitive Applications knew about interactive information systems, and the Mac in particular.

A dummy display system evolved alongside a workbench which gave museum staff the tools to write text and mark cross-references using their own word processors. The system itself automatically formats each screen to agreed visual standards, and creates the hypertext links between the cross-references.

The system was installed a few days before the museum opened, in July 1989: Cognitive Applications estimate that the project took 40 days of the museum staff's time, and 55 days of its own; excluding the hardware, its development cost less than £20,000.

Cognitive Applications see the project as a 'publishing venture' and compare the development of the hardware, housing and workbench to the launch of a new magazine: once the format was established, museum staff could continue to edit and publish the database by themselves.

The curators now maintain the database without a computer specialist in-house. In the event, updating proved to depend as much on internal organization as software skills – an experience common to many of the installations described here. Whether the publishing venture is destined for the page or the screen, the actual work of collecting information and preparing text and pictures commonly proves more demanding than any of the requisite production skills.

The editorial workbench also provides a model for temporary exhibitions, the first of which was prepared to complement *Sport 90*. This provides a quick and economical option for staff, and has the advantage of using hardware and protocols that visitors may already have used, or will soon see again, in the *Study Collection Database*.

Funding for a published version of the database fell through, but the growing popularity of the Macintosh, particularly in schools and design studios, still offers an excellent commercial opportunity for publishing on disk in the future.

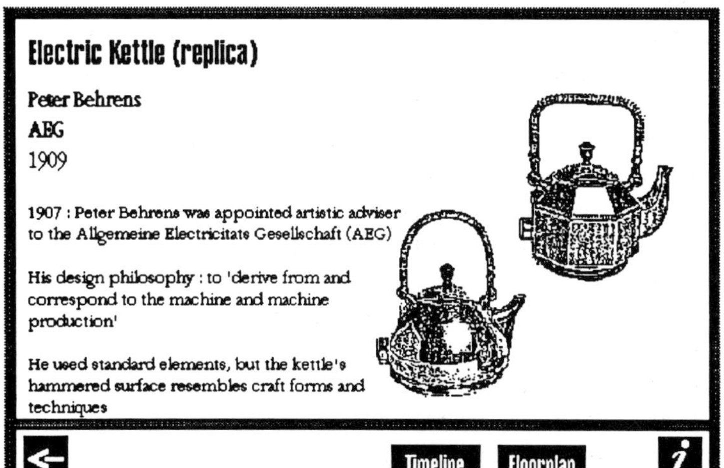

From this screen, users can 'click' on PETER BEHRENS to learn more about this designer, or AEG to see other products from the same company. TIMELINE places the product in chronological order amid other items in the collection, and FLOORPLAN leads to the object itself.

IMPERIAL WAR MUSEUM, LONDON

The Imperial War Museum in Lambeth, south London, uses IV variously to support an audio-visual database of military machines in action, archival footage from two world wars, and images from its own art collection. The museum also features two 'experiences' which simulate moments from life in the trenches and the London Blitz.

Random-access databases of *Aircraft*, *Vehicles* and *Rockets and Missiles* sit amidst the machines of war themselves in the ground floor exhibition hall. Exhibition Officer Penny Ritchie Calder observes that, large as they are, the galleries cannot accommodate tanks in motion, and the machines now carefully maintained as museum artifacts looked much different in combat. The IV systems provide still and moving pictures of equipment in action, with text and graphics to illustrate dimensions, crew positions and other data.

IV also supplies archival sound, film and photographs, together with computer graphics of maps and diagrams, on themes including *The Western Front*, *The Home Front* and *The War Against Turkey* in the World War I galleries, and *The Battle of Britain, The Battle of the Atlantic* and *The Eastern Front* in the World War II galleries.

The longest of the packages so far, *The Western Front*, lasts 25 minutes; most are 15 minutes or so. *The Battle of Britain* is also displayed at the museum's aircraft collection in Duxford, and *The Battle of the Atlantic* aboard *HMS Belfast* in the River Thames. *The Home Front* reflects a trend observed in other projects discussed here, in helping visitors to imagine what life might have been like for them were they, for example, just 18 in 1914. A touchscreen map identifies the local regiments which boys might have joined, or the factories where girls might have worked, anywhere in the country.

The first of a projected series of archival LaserDiscs catalogues 800 images from the museum's art collections, with a retrieval system which allows users to search by criteria such as artist or category.

The AV network is centrally controlled through a system which logs the use of each system on a hard disk. Ritchie Calder notes that the provision of cabling from the control room to the terminals was an integral part of the museum's refurbishment.

Interactive Television provided technical consultancy and services, and Triangle Two, with Flashback, produced the programmes. The systems cost about £12,000 each, with another £8,000 of monitoring equipment and spares in the control room. The first programmes were produced in 1989 for direct costs of £40,000 to £60,000 each (that is, with no provision for the cost of staff time in research, writing and admin, or the use of the museum's own archival material).

Ritchie Calder says that observation of the first programmes led to refinement in the next; for example, the museum found that users wanted immediate action, and were impatient of having to view any introductory material, however brief.

Elsewhere, *The Blitz Experience* uses EPROM chips to control an eight-minute sound and light show which takes visitors from a quaking basement 'shelter' to the smoking ruins of a bombed-out street. It was the museum's first 'experience', and observation has led to some modifications, already implemented in their second, *The Trench Experience.*

The Blitz Experience originally charged a fee on top of the museum's ordinary entrance charge, and had fixed performances at 10-minute intervals. The fee was dropped, and the exhibit will be converted to a walk-through, like *The Trench Experience*, which opened a year later with 8000 hours of development work and the benefit of hindsight.

The Blitz Experience originally used a recording of chirpy Cockneys defying fate against a background of sound effects. This was dropped in favour of an interesting variation on a theme employed, for example, at the Museum of the Moving Image (MOMI). There, actors dress as Russian revolutionaries or cinema cashiers to introduce agitprop or the golden age of Hollywood. In *The Blitz Experience*, actors improvise from rough scripts to provide live voice-overs which can be tailored for visiting groups such as school parties. It proved challenging to separate voices and effects on the original audio recording (prepared by an external supplier), so some few clips of that soundtrack remain, as cues around which the new actors pace their dialogue.

The interactive databases are mounted in sturdy low units with 16" touchscreens at a height accessible to the Museum's many young visitors, and 25" passive monitors above. They can attract such crowds that the museum learned to allow more space around subsequent installations. Much research contributed to 'finding the optimum height and angle of the touchscreens and slave monitors', but it was interesting at the press launch of the new museum to see a lively little boy doing handstands on the touchscreen.

JERSEY MUSEUM

The opening of the new Jersey Museum, in March 1992, launched what could be the largest and most comprehensive interactive display yet produced. The Jersey Museums Service, together with Educational Television Network, used DVI technology to capture some 1000 photographic images, over 30 minutes of film and video, and another 1500 screens of high-quality text and graphics in a presentation which could run for 8.5 hours without interruption, from a single CD-ROM.

The Museum may not want everyone to spend an entire day in front of the screen, but they do hope that the massive archive will provide a service both to visitors and residents of the Channel Island. Tourism is valuable to Jersey, and the museum's Jonathan Carter recognizes that 'the museum has an important role as the "Gateway to the Island", to get people out and about, and looking around. The interactive display can help to link collections in the museum to sites on the island, and ensure that the museum visit isn't isolated from the rest of the time spent in Jersey.'

The museum previously occupied a house of the 1820s, which demanded the trade-offs, familiar to many exhibition designers, inherent in trying to mount public displays in a building of intrinsic historical interest. The new museum is a purpose-built, three-storey facility, in which *The Story of Jersey* occupies one floor, below the art gallery and above the AV theatre, temporary exhibition space and entrance hall.

The package includes information on places of interest around the island, objects in the museum and its reserve collection, fine art, photographs and film, some of which was shot specially for the new presentation. Two interactive terminals each support a larger passive monitor, which allows other visitors to see what those at the touchscreens are discovering. Another, smaller interactive system on the ground floor displays a number of maps in a brief introductory presentation.

Carter hopes that the exhibition will encourage visitors to explore the island and appreciate its rich history and character. But the Jersey Museum also plays an important role in preserving and interpreting the history of the island. In a separate project, an archive of over 30,000 photographs is being captured and catalogued using an image base product called Picture Cardbox. The museum also has a unique collection of amateur and professional film and video, the Jersey Film Archive, assembled in the late 1980s with the assistance of Educational Television Network.

Jonathan Carter feels that multimedia provides a unique tool for making moving footage more accessible to museum visitors. 'Film is hard to use in exhibitions', he says, 'except in a tape loop. Interactive displays provide a very useful way of displaying that material in an interpretive way, rather than just having a telly flickering away in a corner.'

The challenge of vetting this diverse AV collection for material of appropriate quality for transfer to the new medium was relatively easy in that the same supplier worked on both projects within a period of two years. 'We knew what was there, it was very well catalogued', says Alastair Layzel of Educational Television Network.

The photographic archive represents some 1.4 gigabytes of data – enough to fill 750 high-density floppy disks. However fantastic this may sound by comparison to the average home computer, it is not at all improbable by the standards of professional and scientific data storage, for the whole image base will be stored on a daisychain of high-capacity hard disks.

In fact, the museum's design consultant, Colin Milnes, is confident that the new multimedia platforms will be more powerful and durable than IV ever was. He incorporated IV into a travelling exhibition some years ago, and reported that the system performed well over a national tour of some 16 months, but broke down subsequently, when it was installed in a permanent exhibition. He believes that the current generation of hardware and storage media will prove even more resilient.

Jonathan Carter is most enthusiastic about the role of multimedia in bringing large reserve collections to the public in a way unprecedented by any other form of display or cataloguing. 'I think it's going to be the big solution to the reserve collections', he says, 'we can put the entire collection on the screen. When people want to look at photos now, it can take 30 minutes to find and sort the prints by hand; and, over time, the prints can be damaged from handling. With an image base, people can just leaf through the whole collection at their leisure, without handling the originals. No museum should be without one.'

MANCHESTER MUSEUM

The Manchester Museum of Science and Technology, with its many interactive displays, was Britain's Museum of the Year in 1990.

The Space Gallery features two dynamic interactive exhibits developed by Xebec Multi Media Solutions. *The Good Planet Guide* provides a database on the solar system illustrated with digitized photographs from the American space authority, NASA, and LaserDisc. *Journey to the Planets* is an adventure game housed in a darkened space capsule, which introduces material from the National Curriculum while taking children on a treasure hunt through the solar system. It combines more digitized photos from NASA, computer graphics, 3D animation, sound effects and, inevitably, music from Holst's suite, *The Planets*.

The Museum also features an interactive information terminal developed for the Energy Technology Support Unit (ETSU) of the UK Department of Energy. ETSU, based at Harwell Laboratory, produces a range of literature, video, models and other promotional material. The Unit began to explore the potential of multimedia in 1990, and commissioned *Renewable Energy* for use in touring exhibitions and museums.

The programme combines computer text and graphics, digitized photos, animation and digital audio. The opening sequence illustrates the five main topics in the programme as pieces in a jigsaw, which form a complete picture at the end of the introduction. Users can choose an overview of the whole package, any of the main categories, or a game. Each of the main categories has a sub-menu of related topics, and its own game or activity. (*Water Power*, for example, leads to segments on *Wave Power, Tidal Power, Small-Scale Hydroelectric Power*, and a short game called *Hunt for Hydro*.)

The individual topics in each sub-menu are presented with a brief overview (using digitized images, audio and captions) and a database, which visitors can interrogate by choosing questions from a list. These are categorized by Who, What, Where, When and How, and represent the questions most frequently asked about the various topics.

The package is wholly contained on the 640MB hard disk of a PC-compatible Dell computer, with a Soundblaster Pro audio card and SuperVGA graphics with resolution of 640x480 in 256 colours. It now uses a touchscreen, but could employ a mouse or trackerball. The design of the purpose-built housing, with its vivid neon signage, was a critical part of the specification, for the system had to be sturdy and tamper-proof, and yet inviting to a wide audience with virtually every level of interest and ability.

Mick Landmann of Vivid Interactive, who helped to develop the package (which was produced by MLS), observed that 'careful attention must be paid to every single discrete element – text, audio (spoken word, music, sound effects), graphics, animation, pictures. These all have to be separately and expertly produced, then integrated into the whole, which is a specialism in its own right.' The integration of the various elements to ensure a seamless flow of data through the system proved to be a critical aspect of the project, requiring the skills of a computer programmer for every minute adjustment.

Landmann notes that new areas of expertise are developing alongside technical innovations. When this package was produced, for example, the use of digital audio from the computer's hard disk was relatively novel, and there were fewer audio specialists than graphic designers or animators.

As he estimates that '90% of the package is spoken word supported by pictures, graphics and animation', it was important to hire specialists who understood audio techniques, and could help the multimedia team enhance their own skills. Developing such new specialisms is, he says, not only a question of mastering the technology, but also considering wider issues such as acoustics and the design of housings.

The computer also collects data about the visitors and the segments they choose, which can be supplied on a print-out at any time. The housing conceals a printer, as well as supplies including printer rolls and ribbons, screen cleaner and a cloth.

The facility to collect and analyse data was used in two rounds of evaluation at the Department of Energy's displays at the Middlesex Show and the Green Show in Bristol. Independent researchers also interviewed a selection of visitors, whose comments helped to inform adjustments to the package.

After field testing, a new screen was designed to explain the use of the databases more clearly, and the audio scripts were revised to ensure that no one segment lasts more than 90 seconds. Landmann says: 'In a museum or exhibition environment, the user tends to demand constant interaction. Long passive sections in which the user has no control tend to be off-putting – and users often lose interest.'

MUSÉE CARNAVALET, PARIS

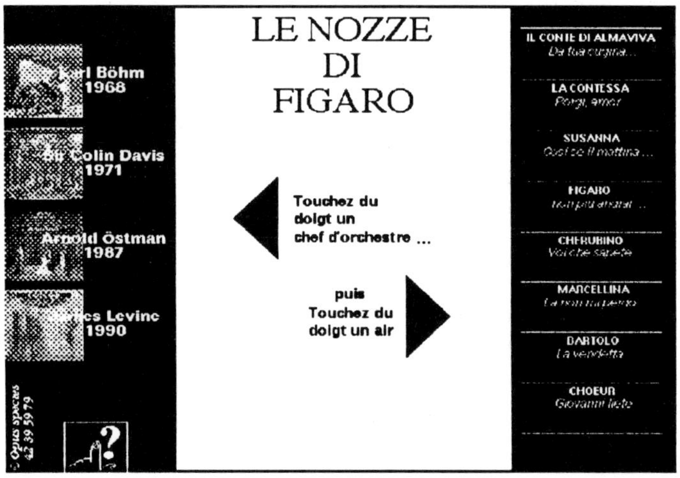

Mozart, of course, achieved multimedia fame in 1991, the bicentenary of his death. *Mozart à Paris*, a temporary exhibition at Musée Carnavalet, combined a CD recording of performances of his music with screens of text and digitized images which explored Mozart's experience in Paris, and the impressions he left on contemporary Parisians.

Four compositions provided the structure for the programme. *The Concerto for Flute and Harp* allowed visitors to compare the same piece with modern instruments and those (original or reconstructed) from Mozart's time. With the touchscreen display, visitors could quickly switch from one version to the other to compare and contrast interpretations. The text screen displayed the score, and followed the audio recording to the very bar to guide viewers through a reading of the notation.

Les Noces de Figaro offered excerpts from four recordings of the opera made between 1968 and 1990 *(above, right)*, to compare the effect of different conductors, orchestras and soloists on the same work, and to study the different types of voice required by the opera's seven principal roles. Listeners could compare the baritone of Fischer-Dieskau as the Count with that of Hermann Prey as Figaro, or the different female timbres required for the roles of the young Cherubino and the mature Marcellina. *Idomeneo* presented the complete libretto of a less familiar opera in French and Italian to help listeners follow the action scene by scene and act by act.

La Symphonie parisienne drew on contemporary letters and journals to create a vivid portrait of Mozart and six Parisians who knew him at the age of 22, when he wrote the Symphony in D on a visit to Paris with his mother. The Parisians included the concert director and first violin at the symphony's first performance; their impressions of Mozart, and his comments about them, were juxtaposed in adjacent text screens (*below, left*).

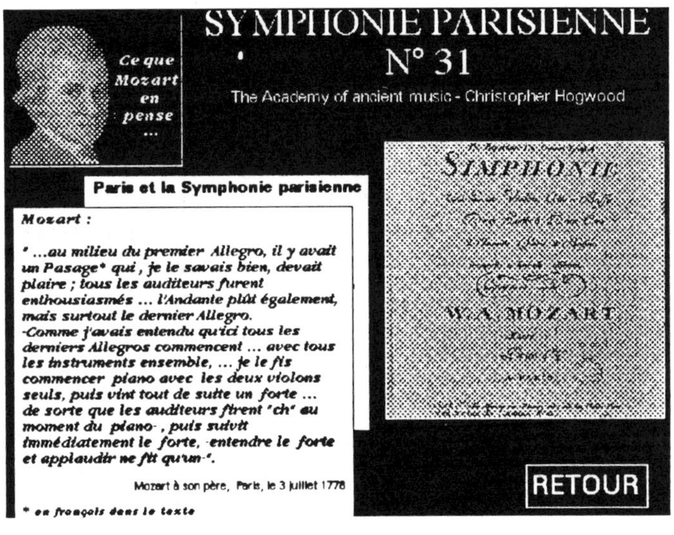

The package was produced by Opus Species, whose recent work also includes a pilot project for the Bibliothèque Nationale which illustrates the use of interactive media with audio collections. A combination of CD-DA, digitized stills and graphics on the Macintosh explores *Les opéras de Georges Bizet*. Extracts from five works are complemented with a discography on screen which identifies performances by different conductors and soloists. *Carmen*, for example, draws on eight recordings made between 1958 and 1989, with a choice of ten excerpts.

Other recent work includes a display for the Post Office, guardians of the Olympic Flame during its journey across France, which was installed at branches in 57 towns and cities along the route. A brief information programme and trivia quiz elucidated the 'fundamental values' of the Olympic Games and the Post Office. Again, the touchscreen system combined digitized stills, original computer graphics, and captions.

Opus Species' Florence Cusson believes that interactive multimedia displays are an important new medium of communication. 'People don't read so much now, but this is a way of getting them to understand the message. They can learn something just by touching the screen and playing with the system.'

MUSEUM OF CIVILIZATION, OTTAWA

The Living Arctic made its debut in an exhibition of historic and modern artifacts from northern Canada at the Museum of Mankind, the ethnography department of the British Museum. A second version was produced for the Children's Museum of the Canadian Museum of Civilization, and a touring exhibition called *Bravo! Canada.*

The Living Arctic uses three touchscreen programmes and a multimedia database to examine traditional society in a changing economic environment. In *Northlands,* players locate the habitats of plants and animals on a map to build a composite picture of the northern landscape; locating a settlement leads to a short video clip of daily life there. *Storyteller* illustrates stories told in English, Dene and Inuit with images from native art.

In *Seasons of the Hunter,* players hunt for game to feed their families and furs to sell. They must choose appropriate supplies and weapons, identify legitimate game, and maintain a balance between hunger and overkill when they hunt. Help is provided by 'consulting the elders', who advise the young to learn by watching and listening. There are glimpses of an igloo, but players soon learn that finance for the hunt now depends on welfare payments and arts and crafts, since regular paid work is scarce.

The project was conceived by Jonathan King, Assistant Keeper at the Museum of Mankind, and Indigenous Survival International. The Canadian government's contribution was coordinated by Dale Thompson, Director of the National Identity Task Force within the Secretary of State's Department (the ministry of arts and culture), and Brian Roberts, Director of Renewable Resources within the Department of Indian and Northern Affairs, with support from the Department of External Affairs.

The IV was made by New Media, in a team led by Nicholas Lewis and Philip Smith, with programming by another London-based company, IE. The project was conceived three years before its debut in London in 1988; Canadian versions were developed for the Museum of Civilization, which opened in 1989, and *Bravo! Canada.*

In this respect, the project is not untypical, in that it ultimately involved a great many people, but was largely driven by a handful of enthusiasts. It is one of the keenest lessons learned in the British IV market, that one visionary can do much to introduce new technology to an organization, but it is important to involve people at all levels – particularly, senior decision-makers – to ensure support for long-term development of both software and systems. (The Imperial War Museum's production team identified 'Script submitted for approval by Director-General and other senior staff' as the sixth discrete stage in their 14-part development plan.)

Dale Thompson was attracted by the quality of New Media's previous work and of British film, video and television production generally. 'You can see it in the ads – British TV is more intellectually engaging', he says. 'We have to recognize that the users' familiarity is with a TV screen, and that we have to engage that audience with a quality product if we want them to get our message. Funding an exhibit in the UK wasn't easy, and working across the North Atlantic sometimes gets difficult, but I think the pay-off, in the greater appreciation of the aboriginal lifestyle, was considerable.'

In fact, the Canadian version reinforces some messages, particularly the danger of overkill. 'Harvesting of wildlife is an integral part of the native tradition, and this programme makes that clear', Thompson says. It was important to the Assembly of First Nations that other cultures should understand the role which hunting plays in the lives of the indigenous people of the north.

Thompson describes *The Living Arctic* as 'adventures in the fur trade', but says 'I refuse to call it a game: it's a very good learning instrument'. He calls the system 'a touch-and-point show ... as interactive technology goes, touchscreen is fairly unsophisticated, but it depends on what your market is. For museums and travelling exhibits, you must approach the end-user as a computer illiterate. If you just want to tell them what bus to take, it can be simple, but if you want to educate them, it's got to be fun, and that takes more than just systems design – it requires quite a degree of artistic creativity.'

The Living Arctic used no original video footage but, like many projects then and since, had to set the saving in production costs against the time and effort required to research and acquire stock footage from five sources in two countries – the Dene Nation, the National Film Board of Canada, Granada Television, Owl Television and Yellowknife Films. Many computer graphics were also specially produced for the three games: New Media's Managing Director, Dick Fletcher, notes that the main screen for one game was created from three separate graphics.

The Canadian version is also bilingual, with a choice of text and commentary in English or French. 'The ability to switch from one language to another is critical', Thompson says, 'but the second language is not a translation *per se*. It's a different version, the two scripts do not directly compare.' This has produced some weak links, he says, which could have been addressed by producing the two versions together: 'The whole package is not as good as it could have been because some visuals don't exactly fit the French language version of the script. Translation should not be an add-on or an afterthought.'

THE CANADIAN MUSEUM OF CIVILIZATION

The Canadian version of *The Living Arctic* is used in the Children's Museum of the Canadian Museum of Civilization in the Ottawa-Hull capital region. The whole museum is cabled for fibre optics and designed to support projection systems which can be adapted well into the next century. Exhibition planner Paul George says that the museum hopes to exploit its telecommunications facilities to provide shortwave station and ham radio link and/or computer networking to other children's museums.

In London, *The Living Arctic* was used by the Museum of Mankind's education service during the week, and put into the exhibition at weekends. In the Children's Museum, it is lodged in a corner of a discovery room alongside an inflatable igloo, an Inuit *Sesame Street*, and a costume collection with a closed-circuit TV loop that allows visitors to see themselves on the big screen. Admission to the Children's Museum is limited to 150 at a time, but the atmosphere is often frenetic. '*The Living Arctic* would be better in a quieter area', George says, but even with dozens of children actively enjoying other exhibits, it seems that some can sit placidly in front of the IV screen, concentrating fully on the programme there.

The system is housed in a plain box at a height accessible to children seated on large beanbags in front of the unit. Dale Thompson reports that, 'The infra-red screen is really tough: you'd have to take a pocket knife to the membrane to damage it'.

This idea has not yet occurred to any visitor, but staff recognize that the hands-on approach favoured by children's museums like this one does subject the exhibits to considerable wear and tear. Paul George also notes that visiting families tend to control their children more strictly in other parts of the museum, 'whereas the atmosphere in the Children's Museum – a space designed specifically for this young audience – encourages parents to give their children the freedom to explore the many interactive and participative displays. Some parents initially think that there is more play than education – at least until they begin to see how much children learn through play.'

George observes that 'parents like to interact with their kids, but they don't want to be embarrassed by their own lack of computer skills. Although children have no hesitation in sitting down in front of a computer screen and learning how to use a given programme, instructional signage would help to attract less computer-literate adult visitors to the interactive display and encourage them to take the first step.'

The Children's Museum also uses LaserDisc for AV in conventional displays. 'We'd be having to repress tapes every week or two otherwise'.

In rather quieter environs, the Museum of Civilization's History Hall features a database, *500 Images of Colonial Life in Upper Canada*. It combines icons and text in a choice of English or French on computer screen, with stills on a video monitor. The text initially provides basic information to place each image in context and help users to identify and compare features of interest. An option to 'read about this item' presents a fuller description. The housing is, again, utilitarian, and designed for use by an adult standing in front of the unit.

BRAVO! CANADA

The Canadian Secretary of State's Department sponsors exhibits at four major exhibitions and 25 regional fairs each year. It also supports the *Bravo! Canada* 'museum mobile', a travelling roadshow housed in three 23m/64' tractor-trailers.

Bravo! Canada undertakes a national tour from April to December each year to towns and cities which do not have museums. The exhibits usually spend five days at each site: Dale Thompson described the tour as 'virtually a transport operation – we've even had these things barging up the Mackenzie River to communities of 400 and 500 people'. *The Living Arctic* tours northern Canada as part of the *Bravo! Canada* show, in a rare instance of an anthropological study's actually being shown to the people whose lives it describes. Over two million northerners visited *Bravo! Canada* in first two years that *The Living Arctic* was on tour.

The exhibits do not travel with a technician so 'they must be designed to be cared for by people – hired locally, on short contracts – with little or non-existent computer skills', Thompson says. 'It has to be kept to the level of "you turn on the computer by doing this ..." .' The system did fail once, and a new LaserDisc player was shipped north to revive it, but Thompson is satisfied that the technology is robust enough for any museum application.

His message to other institutions is clear: 'The IV package is the most worthwhile part of the exhibit ... it's a drawing card, it really is. We have achieved enormous popularity: it's a big winner. You should be going this route.'

MUSÉUM D'HISTOIRE NATURELLE, PARIS

On a marché sur la Terre (*Life on Earth*) was a temporary exhibition for the Natural History Museum which ran at the Jardin des Plantes de Paris from May 1991 to January 1992. An important new gallery is scheduled to open in 1993; in the interim, temporary exhibitions are providing insights into the collection and previews of the displays which the new installation may feature.

In this exhibition (as in that for Taiwan's Museum of Natural History), interactive information terminals complement dioramas of specimens from the collection. Three separate systems (with some 30 entries in each) describe mammals, reptiles and birds. This is relatively small as a database, but simply reflects the number of objects actually on display.

With so few items in each interactive collection, each menu can be graphically represented on one screen *(above, right)*. Clear line drawings illustrate the profiles which each specimen presents in the diorama. Visitors simply touch the drawing to see a more detailed image and read a short text entry about the species (*below*). As the information appears on the screen, a spotlight illuminates the specimen in the diorama.

The information packages were produced and delivered on Apple Macintosh computers. There is no moving footage, and the only audio supplies samples of bird calls and song.

The multimedia packages were produced by Opus Species of Paris, who have cornered the market for European ornithology on CD. Their *AVES* multimedia encyclopædia on European birds is published on CD-ROM by CEDROM Technologies and Sitelle at 2490FF per disc. The package was developed with support from the Ministry of Research and Technology.

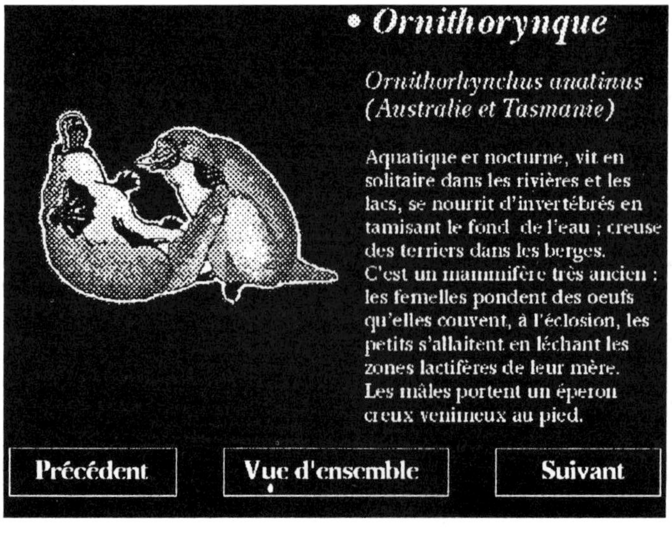

The first volume concentrates on birds of France, with French text only, although all European species are represented. The second covers water birds and the third studies natives of mountain and Mediterranean environments; English-language versions are in preparation. The packages are compatible with both PC and Macintosh computers.

Each database includes up to four colour illustrations of each species (to provide comparisons of male and female plumage, and silhouettes of birds in flight and at rest). Colour-coded maps illustrate the native environs, and winter and summer homes of migrating birds. Audio clips provide samples of individual calls and songs, and the medleys which are most likely to greet the nature-loving ear. The first volume of *AVES* comprises 280 drawings, 97 maps and an hour's worth of audio.

The screen displays employ three directional arrows to move back and forth between individual families and species, and move up and down lists of birds in one family, or the index of families. Five icons – a scroll, a book, a telescope, a compass and a singing bird – lead respectively to the index, database, illustrations, maps and audio clips.

Opus Species were also among the first companies to win funding from the Media Investment Club, an initiative of the European Communities' Media 92 programme, to produce *Les Oiseaux d'Europe* on CD-I.

MUSEUM OF NATURAL SCIENCE, TAIWAN

Taiwan's National Museum of Natural Science is an interpretative centre which has proved tremendously popular in a culture which places great emphasis on education (one designer reported that, on an early morning visit, she saw whole families queuing for admission at eight o'clock). Two British multimedia developers have produced exhibits for different parts of the museum.

MINI-THEATRES

Media Projects International (MPI) produced four multimedia AV packages for mini-theatres on the periphery of the museum's main hall, a large rotunda with high ceilings and large windows.

MPI's Rosalie Vicars-Harris believes that 'it is important to destroy the cathedral-like reverence' which intimidates many casual visitors, and feels that part of this museum's success is 'that it doesn't have a collection!'. She notes that low ceilings, carpeted floors, and vibrant use of colour in many areas of this museum help to create a friendly, non-threatening environment. The opportunity to sit down and watch an AV is also a welcome feature lacking in many museums.

The mini-theatres are small cinemas accommodating from 30 to 400 people on short benches which MPI's Malcolm Lewis describes as 'comfortable for up to 15 minutes at a time'. The entrances to the mini-theatres are screened in Perspex with colourful, inviting signage. Lewis observes that this also helps to reduce the conflict in ambient sound between the display halls and mini-theatres.

MPI employed a variety of AV effects and systems, and made extensive use of LaserDisc to store audio, images and control programs. *The Origins of Life* is a nine-minute presentation on the cell which employs four slide projectors and a rear-projection screen, together with two flanking video monitors. *Voices in Nature and Man's World of Sound* compares a variety of sounds from the natural world and human environs. The visual display is supplied by nine projectors on a front-projection screen with raised panels; the three-channel soundtrack is supplied by LaserDisc. *The Geometry of Nature* is a single-screen AV which makes extensive use of computer graphics to illustrate the complex geometry of natural forms and patterns, and the ways in which they can be analysed and employed. *Relativity* combines slide projection and LaserDisc to introduce the basic principles of Einstein's theories.

Malcolm Lewis believes that, in addition to conventional monitors and free-standing terminals, options such as rear-projection screens can improve the visual quality and interest of the displays, and provide designers with facilities to support new exhibitions well into the 1990s. MPI's recent work has extended into the world of animatronics, in a multimedia presentation for a permanent tourist attraction in Sicily, and a dark ride about pirates at Gardaland theme park in northern Italy.

DIORAMAS

The second phase of Taiwan's Natural History Museum will open in 1993. Among its displays will be a number of large dioramas depicting the flora and fauna of the world's ecosystems from the savannah to the mangrove swamp. The sweeping dioramas, with visually complex displays sealed behind glass, will be enhanced for visitors with the aid of a dozen interactive terminals.

The IV packages are being produced by New Media in London. Creative Director Philip Smith describes the video sequences as 'vivid clips on which to base interpretations – where the dioramas are necessarily static, the multimedia displays add sound and movement to complete the picture'. All the programmes in both phases of the museum's development were produced with audio and captions in Mandarin, which posed a certain challenge to both production companies.

New Media, like MPI, has considerable practical experience both of designing displays for brand new buildings, and dealing with clients on other continents. Smith notes that, technology notwithstanding, he has observed a significant change over the years in closer working relationships between exhibition designers and the specialists who supply multimedia and AV displays. He stresses the importance of 'getting in early' to ensure that the interactive displays form an integral part of the new gallery or installation.

This can be critical in the supply of ducting, cabling and other services, and to ensure that ambient noise and lighting satisfy the demands of various displays. Equally, involving external suppliers from the beginning can also contribute greatly to the visual and intellectual quality and consistency of all the exhibits in the gallery. Forward planning with specialists can inform the management of change well into the future.

NATIONAL GALLERY, LONDON

The Sainsbury Wing of the National Gallery opened in June 1991, with one of the most innovative interactive media systems yet seen. The development of the project is a testament of faith in emerging technologies, for neither hardware nor software was commercially viable when the specification was prepared in 1988.

The National Gallery holds the UK's national collection of western European painting from the 13th to 20th centuries. The Micro Gallery occupies a long, narrow room just off the new wing's grand staircase. It provides images and information on over 2000 items in the gallery's collections.

Even with a dozen terminals, the Micro Gallery is usually full to capacity, and visitors are often obliged to peer over others' shoulders while waiting for their turn. Three systems immediately inside the entrance require visitors to stand, a strategy designed to encourage through-put. Seven more are staggered in a row, with stools to accommodate up to three users at each station; these are designed for longer visits, and can be booked in advance. Two further systems near the entrance are positioned for use by visitors in wheelchairs.

Access to the image base follows four main routes. *The Artist* allows users to select any name in the collection to see a brief biography and a 'thumbnail' gallery of miniature reproductions; touching any one retrieves the larger image and catalogue entry. *Picture Types*, similarly, uses the thumbnail gallery to divide the collection into genres such as still life, landscapes or portraits, and thence into manageable categories such as portraits of royalty.

The *Historical Atlas* begins with a map, and allows visitors to refine their search to a specific place and period, to find all the artists and work in the collection which are connected with any one city and era. *General Reference* allows for thematic approaches (for example, to explore themes rather than periods or genres). The quick response times and flexible structure allow users to move rapidly from one part of the package to another, to pursue ideas freely across the various databases.

(The system is, of course, ultimately limited to what its designers thought appropriate to provide: one visitor from another museum was disappointed not to find explanations of the iconography, for example, of saints' emblems and religious imagery.)

The retrieval program uses a combination of visual prompts such as the thumbnail images and maps, and verbal prompts including lists of topics and keywords in text entries. Visitors can learn more about art history terms, techniques and conventions from a menu of keywords which is introduced by a screen of initial letters: if a reference to gouache is puzzling, the alphabet leads to all the entries beginning with G, and thence to a definition.

The objectives of the project were to supply an interactive image base in full colour on large screens, for the use of visitors who would typically range from tourists to art historians, and who might know anything or nothing about computer-based systems. The hardware choice fell on powerful Macintosh computers with large hard disks (1.3GB) and 19" touchscreens.

Text, graphics and animation were assembled by 21st Century Systems, working with gallery staff in a team which included (in addition to curatorial staff) two researchers, an editor, two graphic artists and an image technician. The software was developed by Cognitive Applications, with a combination of programming in the C computer language, and a tailor-made editorial workbench based on Apple's HyperCard. The Micro Gallery is sponsored by the American Express Foundation.

The system holds over 2000 images, scanned from 8"x10" colour transparencies and 35mm slides. The development team experimented with resolution at three levels (24, 16 and 8 bits per pixel), and with such compression systems as were then available. With some customizing, the 8-bit option provided the optimum combination of visual quality, storage capacity and response times. Additional work with custom-made fonts (anti-aliasing, to smooth sharp lines, angles and curves) improved the visual quality of the captions and text displays.

Nine of the workstations include a LaserWriter printer, which provides a free print-out of the user's personalized gallery tour on request. A smart card reader also allows the gallery to administer on-the-spot sales for black-and-white reproductions: visitors pay for the cards in advance, and the readers deduct the cost of each print from that credit limit (currently, at a rate of £1 per print). Alex Morrison of Cognitive Applications says that 'this is the first project to take printing seriously'. There have been predictable problems with paper jams and toner supplies, but 'visitors are spending serious amounts of time and money' on the print-outs, and the experiment has proved highly successful.

CASE STUDY

NATIONAL MUSEUM OF RACING, USA

This case study was written for the NIVC journal Interactive Update *in 1989 by Neil Sandford, who was then the Managing Director of* The Soft Option. *It describes an installation at the American National Museum of [Horse] Racing in Saratoga Springs, New York, which was completed in July 1988.*

Some aspects of this project are unique but, as a candid record of the many twists and turns which occur in virtually any project, this provides as good a case study as any in the present market.

There have, of course, been significant developments in the market for interactive multimedia systems since this project was completed: the same installation, conceived five years later, would probably have used different systems. (However, it is interesting simply to consider how recently the current range of multimedia platforms has developed, and how profoundly any installation may be influenced by the choice of media available at the time of its inception.) The Soft Option itself is no more, but many of the people and companies cited here are still active in the multimedia market.

On 14 July 1988, at five o'clock in the afternoon, the American National Museum of Racing and Hall of Fame re-opened after a two-year refurbishment.

For The Soft Option, it was the culmination of five months of hectic development work. For ex-pirate radio DJ Peter Brady (by then the eponymous head of Pete Brady Communications), and project manager and designer Ivor Heal, it was the result of two years of commitment to a project that had started, back in the summer of 1986, with no more than an agreement that the museum would contain an 'interactive' exhibit.

In the event, the final result was far removed from the initial concept of a 125-screen videowall ...

BACKGROUND

In the USA, race meetings tend to be occasional but lengthy affairs. Racetracks open for a season, and receive large numbers of out-of-town visitors. The National Museum of Racing is based at a racetrack in a major tourist resort, Saratoga Springs (celebrated in song in *Guys and Dolls*). The museum was planned as a tourist attraction in its own right, as well as a centre of educational and historical value.

The Hall of Fame, as its name suggests, is a memorial to the famous names in American horse racing. Each year, yet more famous horses, jockeys and trainers join this distinguished group: the objective of the interactive programme was to give visitors easy access to accurate and consistent information about all the 122 horses, 59 trainers and 64 jockeys already celebrated – and about some of those who would be inducted in years to come.

Peter Brady now claims to have had no previous exposure to IV beyond displays at Disneyworld and EPCOT, and occasional magazine articles; however, he recognized that the large amount of information to be accessed and the need to update the programme annually both legislated in favour of a computer-based solution.

As well as the IV-based information booths, the Hall of Fame contains a permanent exhibition of racing silks and plaques commemorating the inductees (what other common noun describes the two- and four-legged stars of horse racing?). It is also the venue of a 15-minute AV spectacular called *Race America*, which runs every hour on the hour, requiring the IV screens to fade and curtains to close across the cabinets of racing silks. Apparently someone drew the line at under-floor loudspeakers to reproduce the thunder of galloping hooves, but that was the only constraint on the museum's showpiece.

RESEARCH

The decision to use a British design team for the museum was justified by the 'track' record of Ivor Heal and editorial consultants Tempus, who had completed a similar project at Newmarket. Logistics and cost provided good arguments for bringing the construction of exhibits and video post-production to the UK as well. Contracts were exchanged early in 1987, and the design of the interactive exhibit was underway. One of the first tasks was to identify potential sources of live action and stills: Peter Brady and Pat Conner of Tempus spent much of the next year tracing suitable material.

They quickly discovered that there was no convenient central archive, not the least because racing is not as frequently televised in the US as in the UK. Eventually, after a few hiccoughs, material was gathered from America's 40 major racetracks and the racing fraternity itself. Racing and bloodstock magazines proved invaluable in providing contacts with owners and trainers, most of whom were enthusiastic about making their private collections available to the museum.

Altogether, the footsore research team collected and shipped enough still material to fill nearly three full-height filing cabinets back in the UK – 10 drawers' worth of prints, slides and original artwork. Only fragile glass plates – about 30% of the black-and-white material – were transferred locally to tape. Peter also had some 60 hours of live action material, most of which came from patrol cameras at the racecourses, and so comprised one fixed shot from the final bend to the finishing line. One favourite sequence, shot with a hand-cranked camera, proved so exciting for the cameraman that he cranked faster and faster (forgetting that this makes the finished film look slower and slower) until the leaders appear to be crawling toward the finishing line. For one horse, the researchers could find footage of only one major race, which the fleet-footed quadruped won by such a distance (40 lengths) that it was off camera at the finish!

Copyright fees ruled out the use of much modern material; apart from footage provided by the racetracks, most of the live action came from films commissioned by the owners or from early TV footage. To the consternation of at least one facilities house, some of this was on volatile acetate stock. When asked whether insurance had been arranged for the journey of this irreplaceable material across the Atlantic and back, Peter replied, 'Through my broker' – the sum assured was $10 million.

PRODUCTION

The editing process began in the UK in November 1987, with all the material safely on Betacam. The plan was to have one programme running on all six booths in the Hall of Fame. This would require dual-disc players, and great discipline in selecting material. However, early in 1988, it was decided to cover all 40 major racetracks as well as the 122 horses and 123 bipeds in the Hall of Fame on the disc. This proved too much even for two sides of PAL LaserVision (PAL having been chosen over NTSC precisely because it offered a critical few minutes more of running time). The solution was to spread the material alphabetically over three videodisc sides, and pair off the booths so that each of the three discs sides would run in two booths at a time.

Meanwhile, the design of the rest of the museum kept Ivor Heal's team occupied – and requests trickled in for AV material for those galleries as well.

The museum features large fibre-optic displays in two galleries. One, *Blueprint for a Champion*, explains the physiology ('conformation') of a thoroughbred racehorse. The gallery's fibre-optic panels and large video monitor 'wake up' when a visitor enters, and present an illustrated commentary on the anatomical similarities and differences between horses and human beings.

(The horse's ankle is actually half-way up its leg, so a racehorse spends its entire working life pounding along a continuous left-handed bend on four toenails! This and other facts of equine life were provided by an academic authority who was also a racing fanatic. He was recorded in a broom cupboard under a staircase during the research trip, but his unscripted presentation carried such warmth and feeling for the subject that complementary visual material was subsequently edited to fit his narration, which was transferred directly from the original tape, complete with footsteps on the stairs above.)

SETBACK

For a large and complex project (the refurbishment as a whole cost several million dollars), things had been proceeding remarkably smoothly when, early in 1988, a critical member of the team – the interactive video specialist – simply disappeared, never to be heard from again (not, dear punters, a common occurrence in this otherwise reliable industry).

Thus, The Soft Option first heard of the project on 15 February 1988, when I was asked by a client to contact a friend of his who seemed to be in a bit of a spot. That evening, Peter showed me a rough edit of *Blueprint for a Champion* and presented a list of the equipment and software requirements for which he no longer had a supplier.

This included several interesting points. The schedule required that all the hardware for four galleries' worth of exhibits be ready to travel by sea to the USA on March 25, for arrival in May. On the basis of previous experience, it had been decided that touchscreens were unsuitable, and that a keypad was required; this would eventually take 12 weeks to manufacture, but had not yet been designed. The whole museum was to run from a single power switch, and each exhibit was to perform a self-diagnosis every morning (and either report an error or else start running): this meant that even the simplest AV presentations needed a microprocessor controller. The Hall of Fame systems also had to close down once an hour and wait for the big movie to finish on the main screen. Then there were the fibre-optic panels! Nobody quite knew how these would behave, and there was no chance to find out until everything was actually installed in the museum. (Fortunately, the decision to use a microprocessor controller allowed us to make on-site adjustments to the pauses between each section of dialogue, to compensate for the response time of the hardware – and the viewer.)

Needless to say, we accepted the challenge and started work on producing a set of specifications.

SYSTEM DESIGN

By far the most important aspect of the various interactive exhibits in the museum was the ease with which people learnt how to operate them. The conformation sequence started automatically whenever somebody entered the empty room: that was no problem. The genæalogy used a fibre-optic wheel to trace the bloodlines of individual horses; the instruction TOUCH THE NAME OF A HORSE TO SEE ITS LINEAGE was blazoned across the 2mx1.5m photograph that formed the basis of the display. An A4 concept keyboard was built into the housing, with a silk-screened overlay giving the names of ten champion horses. It worked.

The Hall of Fame provided the most design challenges. As well as selecting from some 175 entries, visitors had to be able to break out of the attractor sequence, start and stop the moving sequences, call up supplementary pages of information and get back to a menu from any point. The design solution was to provide back-lit icons on the keypad presenting the four thematic areas (horses, jockeys, trainers and tracks), as well as a means to select from lists in each category, using three big illuminated red buttons labelled START, NEXT and CANCEL.

The first challenge was that the final design for the keypad had to be submitted to the manufacturer before the actual selection method had been determined. One option was to reduce the list to a single screen display by entering only the first two letters of the required name. The knowledgeable user could probably manage this, but the casual visitor might find it a bit daunting The alternative was to present a page of information, from which a single keypress makes the selection.

As NEXT was already in the syntax for the system, it was felt that users could easily use this key to page through a list of names. To leave the options open, keys marked A to Z were provided and, in the final programme, individual pages were restricted to a maximum of 26 names.

Of the three keys, START activates a clip of moving footage, and is only lit when this option is available (the word is reproduced on the screen as a reminder); this button is also used to break out of the attractor loop at the beginning of the programme. NEXT is always used to move forward: the user is introduced to this button after the explanatory text screen that follows the attractor. CANCEL does just that, and takes the user out of any part of the programme back to the main menu.

There is no method of navigating up, down and sideways across the menus - there is no need to. There are no more than three pages of names in each thematic area, so any entry can be reached by a minimum of two and a maximum of four keypresses.

In practice, the keypad worked well. The visitor is first presented with a video attractor that includes a round red computer graphic labelled START. The visitor presses this: nothing happens. The visitor should then look down and see that the START button on the keypad is the only one lit up, and press it. The screen then changes and the visitor gets just one screen of explanatory information, which includes graphics illustrating the four thematic icons, and a reminder that only entries from A to F (or whatever) appear in this booth. The visitor is then invited to press NEXT, and begin to explore the programme.

OUTCOME

The museum opened on time and within budget but, for the IV system, it was a photo-finish. The software was updated within the first year, as new inductees swelled the numbers in the Hall of Fame. As astute racing fans, Peter and Pat anticipated some of these nominations, and included appropriate visual material in the first pressing of the videodiscs. Until the next pressing, new inductees for whom there is no video material are represented by computer-generated entries over a blank screen.

So, what conclusions can we draw? It was certainly cheaper to edit the visual material in the UK than it would have been in the US. There have been some comments about flicker on the screens (and these from pundits rather than punters), but the decision to stay in PAL was justified as the quantity of still and moving footage would not have fitted the NTSC discs. The only shortcoming was that very few people were actually observed using the NEXT key to page from one screenful of names to another: the upgrade added the NEXT icon to the screen display as well.

It was fun – most of the time. (A few moments that weren't fun: ringing up one manufacturer for a delivery date and hearing that the factory was about to close for its annual week's holiday. Putting a brand new employee straight on a plane for New York with assurances that things wouldn't always be like this. Engineers sitting idle in Saratoga Springs while US Customs took their time over hardware lodged at JFK International. 'Same day delivery' that took 72 hours. The near-asphyxiation of the engineer who tried to work in the Hall of Fame on the day that they glued the carpet down. But that's another story, or several, none of which has to do with the technology.)

The museum was hit by lightning shortly after it opened, but even that caused only minor problems. We received a great deal of support from our hardware manufacturers – Visage, Sony, Hantarex and Qubie – all of whom provided warranty cover in the US for equipment purchased in the UK.

Would we do it again? Of course we would – but next time, isn't there somebody a bit closer than Saratoga Springs who needs to put several thousand pages of information into a display area four feet wide?

NATIONAL WATERWAYS MUSEUM, UK

The National Waterways Museum in Gloucester features three interactive presentations on aspects of the history and use of the nation's inland waterways. The museum is housed in a former warehouse in what were Gloucester's own working docks. Its design leads visitors on a labyrinthine path through many colourful displays.

Canal Builder uses video footage of construction features such as embankments, locks and tunnels to help the user design a canal through changing landscapes on the touchscreen. Design challenges include a hill, a valley and a gradual incline, which must be accommodated to provide a smooth and safe journey for the long, narrow canal boat.

Operating a Lock also uses a touchscreen to simulate opening and closing lock paddles and gates. The consequences of error are vividly illustrated, but players can try again until they manage to guide the boat safely through the lock.

Steering a Boat uses video shot from the helmsman's perspective and projected onto a large screen to create a life-size simulation of inland navigation. Users stand a few paces from the screen, and operate a real tiller to see the effect of their actions on the course of the boat.

The three packages were originally developed by 3T Productions in 1989, with some additional work subsequently on *Steering a Boat* by Xebec Multi Media Solutions. The museum's Des Fforde says that 'the public get a lot out of them – when it's busy, you tend to get a bit of a queue'. The combination of active terminals with larger passive monitors, and the large screen of *Steering a Boat*, at least allows some visitors to watch displays in progress.

There have been some technical problems. On the very day of this writer's visit to the museum, soon after the systems were installed, one display was locked on screen, another was represented by a gaping hole where the hardware had been, and *Steering a Boat* was running as linear video because of synchronization problems with the large-screen projector.

Nonetheless, even a large notice, apologizing for the hardware problems on *Steering a Boat*, did not deter a group of three elderly visitors from experimenting gamely with the rudder, which continued to move freely although it effected no change on the screen. It is observable that, where there are objects to handle, museum visitors seem increasingly confident about laying hands on, if only to 'see what it feels like' or 'what *might* happen'.

Des Fforde says that 'there was a long fight with hardware problems to get them all working properly, but the hardware has been modified and they have been running well for over a year now'.

NATIONAL MUSEUM OF PHOTOGRAPHY, FILM AND TELEVISION, BRADFORD

Martin Oliver at the National Museum of Photography, Film and Television in Bradford reported problems with the early LaserVision systems there, which were similar to those of the National Waterways Museum, and of the pioneers interviewed for the *Videodisc Monitor's* report quoted on page 71.

Oliver estimated that his team replaced burnt out motors at least every two years, and annually on the oldest players. He also had cause to suspect that some equipment donated by corporate sponsors had been used in demonstrations and exhibitions before it was given to the museum.

By way of comparison, the Science Museum in South Kensington, another pioneer in the use of LaserVision for gallery displays, reported that its staff replace motors on its early videodisc players as a matter of routine, before they burn out.

In Bradford, eight copies of the same disc supply still and moving pictures to exhibits throughout the museum. Closed circuit cameras and pre-recorded video support participative demonstrations of TV production techniques such as multi-camera shoots, vision mixing, telecine transfer and chromakey, and let visitors play at reading the evening news.

The technical team's self-taught programming skills produced some special effects, including a screen at the end of a long hall which projects live video 'snapshots' of visitors as they walk toward a closed circuit camera. 'You can do a lot with software in-house', Oliver says, 'so long as you don't get carried away'.

From his experience as a technician and a software designer, he emphasizes the importance of documenting every purchase, repair and development in the life of machine or system. 'You have to know where stuff's from, and where the wires are', he says. He also advises designers to think carefully about the technical support of touring exhibitions, to help local staff support the visiting show.

NATURAL HISTORY MUSEUM, LONDON

The Natural History Museum in London's South Kensington museum complex features many high tech displays, including an interactive videowall on the lifecycle of the carbon atom and an electronic variation on flashcards that tests visitors' short-term memory (*The Carbon Cycle* and *Super Memory,* both made by 3T Productions), as well as animatronic dinosaurs and a complementary information programme on Apple Macintosh. The museum's computer collection comprises an historically interesting combination of PCs, Ataris and even BBC Micros which Ivan White of the Special Effects Section says are gradually being standardized, if only to simplify maintenance and stock control of replacement parts.

EYEWITNESS TESTIMONY

One of the museum's most challenging exhibits demonstrates the effects of stress on memory. Users sit down and touch the screen to start the video. It begins with a leisurely stroll down a city street, shot from adult eye-level. Without warning, there is a burst of violence, a businessman's briefcase is snatched, and the viewer is momentarily threatened with a knife before the assailants escape. Another eyewitness anxiously asks what happened.

Players are then interviewed about details of the incident, and asked to identify a suspect from a police line-up. Theories of memory and recall are demonstrated by various questioning techniques, including visual prompts (identifying the assailant's weapon from a choice of knives) and leading questions (being asked to confirm the colour of one robber's sweatshirt, rather than describe what he was wearing).

The programme was conceived and researched by the museum's Roger Whiteway, and produced by the Moving Picture Company (MPC) in 1989. The video was shot on a single day, but the deceptively casual opening sequence alone required no less than 35 takes, as it had to be shot in one unedited sequence, at eye-level, with a hand-held SteadiCam. Casting required six actors with striking resemblances to one another for the final line-up of suspects, and details about the get-away car had to be scripted after the event as the hire company could not guarantee delivery of any one colour or model.

The rough off-line edit was transferred to write-once Recordable LaserVision (RLV). This allowed the museum's evaluation officer, Katie Edwards, to experiment with a preliminary version of the software which included a choice of questions for certain critical scenes. Data retrieval software helped to collect and analyse cumulative responses to each, to determine which most successfully met the objectives of the programme. From this, the final versions of the video and software were developed.

Even knowing what to expect from the opening sequence, players are unlikely to answer every question correctly and identify the true suspect in just one round. It is observable that visitors will play *Eyewitness Testimony* until they achieve a perfect score – Ivan White admits, 'Once you start playing with it, you do tend to carry on ... '.

ACTION FOR ANIMALS

In 1990, New Media produced a short programme for *Action for Animals,* a special exhibition commemorating the 150th anniversary of the Royal Society for the Protection of Animals (RSPCA). The display combined digitized still pictures, computer text and graphics; the whole package was held on computer disk, under the control of a touchscreen.

The package used a trivia quiz with questions on farming, the countryside and animal rescue to illustrate aspects of the RSPCA's work. Some questions led to straightforward answers, others to statistics or commentary where the issues were less clearcut. (One, for example, asked which of a choice of gin traps, rabbit snares and domestic mousetraps is illegal in Britain. Even visitors who facetiously answered 'mouse traps' were told not only that gin traps are illegal, but that they are often used to trap animals for the fur trade.)

The exhibition also featured linear video and a set of audio case histories presented through telephone receivers – visitors picked up the phone to hear a typical event in the working day of an RSPCA inspector. The various media creatively conveyed fundamental messages about animal welfare, the public's responsibility and the role of the RSPCA.

The whole package was completed in four weeks, although New Media's personal best actually produced a set of interactive videodisc games – complete with customized housing – in three weeks flat, for a special exhibition on Saudi Arabia in 1986. MD Dick Fletcher does not recommend this approach – particularly given the prohibitive cost of express LaserDisc production services – but cites these projects as examples of how quickly simple but functional displays can be achieved.

NORWICH TOURIST INFORMATION

Norwich's Tourist Information Centre uses an interactive terminal to introduce the city's sights in a set of three thematic walks, on *Shopping Lanes and Alleys, Inns and Markets*, and *The City Walls*. There is no print-out, but the Centre sells the three itineraries as illustrated brochures for 75p apiece.

The package runs from the hard disk of an Apple Macintosh, with a colour touchscreen in housing set at a convenient height for children. It makes extensive use of audio and is, wisely, sited well away from the counter where the Centre's staff spend their day.

A map with a view of the town centre provides the starting point for each walk, and the main illustration for the complementary brochures. The principal sights are numbered to illustrate the recommended routes: selecting any number by touching the screen leads to at least one page of text, and at least one illustration. Users can then follow the walk from one screen to the next, or return to the map at any time. The segment on the City Walls uses the map to superimpose an outline of the old boundaries over the modern plan.

Entries for the various sites are uniformly presented with a digitized photo or drawing on the left of the screen, and text on the right. Where modern and historic views are available, a timeline beneath the illustration displays its date, and leads back and forth between modern and archival images (which may be anything from an 18th century print to a photo from the recent past). Arrows flash to indicate when more pages follow in the longer segments.

Icons on the touchscreen control basic functions: help is represented by the familiar italic *i* of an information centre; a pair of boots face right to walk forward, and left to turn back. A little less successfully, a tiny web of black lines represents a map, which leads back to the town plan.

(Even in a programme as well-designed as the Tate Gallery's *Sculpture Interactive*, metaphors such as a movie camera or piece of sculpture proved difficult to convey in a single pictogram. The icon designs provided in toolkits such as Apple's HyperCard are, for their simplicity, models of a specialized discipline.)

The unique feature of this package is that it was designed by local students with the help of the Norfolk Technical and Vocational Educational Initiative (TVEI). The TVEI is a nationwide government scheme 'to create a more highly skilled, competent, effective and enterprising workforce for the 1990s'. This project evolved from contacts between Tim Roderick of the Norfolk TVEI and the City Council and Tourist Agency.

The county's schools are grouped in 'clusters' which, in this case, comprised two high schools and a special school – The Clare School, and Taverham and Hillesdon High Schools. Some 20 pupils, including Sixth Form History students and two A-level Art and Design students, planned and researched the package, which was implemented by Tim Roderick, Steve Lepper and Geoff Rushbrook at the TVEI.

The students were given a list of potential sites as a starting point, although Roderick says that they dismissed some and added others during the course of their research, which was concentrated on one week of diverse activity. This included walking through the town to devise routes and evaluate sites, and poring through local history collections to prepare text and find illustrations.

Given limited resources of time and funds, the TVEI did the work of scanning images and creating software (using Aldus SuperCard), from the students' research and designs on paper. The Art and Design students were given some initial advice from a graphic designer on the principles and conventions of screen design; their contribution formed part of their work for their A-level qualifications, and both have since gone on to art college.

Roderick says that the TVEI unit played an advisory role, and Rushbrook's expertise with SuperCard particularly facilitated the programming, but the pupils made their own decisions: 'We see the strength of the project in the benefit of students working with professionals, as well as enhancing and developing a wide range of skills within an exciting and relevant context.' The structure is occasionally cumbersome – requiring several keypresses to move from one walk to another, or to reach functions such as the choice of languages; however, the package is, above all, consistent in its use of conventions such as icons and colour-coding, which makes it both attractive and manageable.

Versions of the package are being developed for sites in the Castle Museum and Jarrolds, a local department store. Roderick notes that the choice of hardware and production tools facilitates customizing of packages for individual sites – to provide more information about the Jarrold family, for example, or the Castle. There is considerable potential here for the kind of sponsorship on which many such installations now depend.

ROYAL BRITAIN, LONDON

Royal Britain was a tourist attraction in the City of London, which closed in the months between the first and second editions of this report. Nonetheless, the planning, maintenance and design of its interactive systems raise some important points for museums and galleries of all kinds.

Although directly opposite Barbican underground station, which leads many visitors to the Barbican Arts Centre and Museum of London, Royal Britain remained little known and much less visited than either. It charged £3.95 when the Museum of London was free (and, at time of writing, levies admission with a three-month season ticket at £3).

Equally, where the Museum of London is rich in artifacts and splendidly designed, Royal Britain had the feel of a temporary exhibition. It largely comprised portable display panels of the kind used in travelling shows and trade fairs. These were devoted to nuggets of eccentric and often unsubstantiated trivia, gossip and jokes about Britain's monarchs (at least two of whom were credited with 'taking a bath every year – whether he needed it or not!'). The largest collection of objects was devoted to old radios.

Media Projects International (MPI) only became involved with Royal Britain some months after the exhibition opened. Their brief was, effectively, to make it more attractive to visitors who would not be expected to have much formal knowledge of British history. MPI Directors Malcolm Lewis and Rosalie Vicars-Harris observed that, when they first visited Royal Britain, there was no reference to Henry VIII, and very little to Queen Victoria – two of Britain's many royals whom any tourist might immediately recognize.

MPI took the view that most visitors appreciate being led to new ideas from familiar ones, and would expect to see the country's most famous kings and queens before learning more about the less celebrated. They introduced 65 new exhibits and panels throughout the display area – including such obvious topics as a survey of Henry VIII's matrimonial career, which had been neglected hitherto. Several of these drew on computer-controlled videodisc – including a simulation of life before the modern *paparazzi*, with flashing lights and ambient sound, and a clip of Laurence Olivier's *Henry V*.

The original exhibition included a linear videodisc, *The Family Firm*, which presented yet more trivia about the present royal family. MPI were asked to adapt this to the more logical medium of interactive video, so that visitors could use the database creatively on their own. This involved writing a computer program and designing complementary visual material to convert the linear disc to interactivity, and designing a housing. The interactive system was deployed alongside pre-programmed monitors which continued to display the original linear presentation.

MPI had advised pressing a new disc, not the least to include recent events such as Princess Anne's separation and the births of the Yorks' children. Some months later, another company, Xebec, produced a third version of *The Family Firm*. This was wholly computer-based, with digitized still pictures and text. The three generations – linear video, interactive video and computer-based display – eventually ended up in one eccentric installation.

Obviously, any interactive display should consign volatile information to a medium which can be easily updated, and plan the contents of an immutable medium like LaserDisc to ensure that it enjoys a long shelf-life. When Royal Britain was first conceived, in the mid-1980s, the visual quality of LaserDisc had great advantages over computer displays; nonetheless, IV was not then an unknown quantity, and could have supported a versatile presentation that might have been adapted to current events for some years to come.

In the event, two companies with extensive experience in museum and gallery displays could not salvage a project which lacked both the foundation of informed planning, and the support of a coherent development strategy.

Royal Britain also employed two voting systems, one asking whether Richard III arranged for the murder of the Princes in the Tower, the other whether the media treat the modern Royal Family fairly. Visitors were presented with the evidence for and against (which occasioned a substantial display of printed text) and asked to punch keys marked YES or NO to see their votes instantly recorded on a scoreboard.

This is an engaging idea, although a little slack as market research: this writer, for example, voted four times at one station, but not at all at the other. At Royal Britain, even allowing for such deliberate misinformation, the tallies on both suggested sadly low attendances or a reluctance by visitors to participate even in this simple game.

ROYAL MUSEUM OF SCOTLAND

In 1989, the Royal Museum of Scotland in Edinburgh used two information programmes which were produced entirely in-house with a hypertext authoring package called Guide. Both were made in a matter of weeks by Systems Manager Sandy Buchanan and his boss, John Burnett.

Both are clearly enthusiasts with a mission 'to prove to management and curatorial staff that these technologies are robust enough now to put in galleries' – an observation which suggests that there was never any doubt about the efficacy of interactive multimedia, only its physical resilience.

Their team's interest grew out of the Museum's on-going collections management database, which even then held some 250,000 records on a Hewlett Packard minicomputer. As a platform for further development, this proved more useful for curatorial than public information services. The search for a more user-friendly system led to the Human Computer Interface Group at Heriot-Watt University, who suggested hypertext.

A choice soon emerged between Guide, an IBM PC-compatible system developed by Office Workstations Ltd. (OWL) of Edinburgh, and HyperCard for the Apple Macintosh. The Museum already had some PCs, whereas HyperCard on the Mac presented a new learning curve on a system which at that time had less facility for graphics than the PC, so they opted for Guide.

Buchanan estimated that the team probably had £3000-£4000 worth of free consultancy in their various discussions with Heriot-Watt, Hewlett Packard, OWL and others. Hewlett Packard loaned a computer and touchscreen for the initial development work (although the Museum later purchased its own system).

Buchanan observed that, in his experience, most organizations 'will usually do *something* for nothing, if you stick a little label beside it saying KINDLY LOANED BY... You've just got to be cheeky and try it.'

THE WEALTH OF A NATION

In 1989, *The Wealth of a Nation* brought treasures from many Scottish collections to the Royal Museum in Edinburgh. The objects were freely arranged within the exhibition area, so a hypertext system was devised to put the material in context by provenance. The final package described some 160 objects, with another 50 screens of maps and place names.

A hierarchical map-based retrieval system began with a view of all Scotland, asking visitors to touch a region to reach a more detailed map. This led to specific centres of historical and cultural interest. Visitors were asked WHERE DO YOU COME FROM? WHERE ARE YOU GOING? – a tactful choice designed to appeal to natives and tourists.

Visitors could then see digitized still images of locally-made objects from the exhibition and the permanent collection. This proved extremely popular, and excited interest in all manner of artifacts as visitors discovered connections between places they knew and objects in the museum.

The system was indeed robust, and Buchanan reported that the only complaint 'was that there was so much of a queue that it blocked access'. Several of those interviewed here have reported similar dilemmas, and all were ambivalent about an innovation that was almost too successful. Certainly, exhibition designers must not underestimate the effect of popular displays on traffic flow and throughput – nor the frustration of visitors who never get near enough to interact with the system themselves.

DISCOVERY ROOM

The second of the Museum's packages was made for Discovery Room, and after several months of constant use there went on tour as part of the outreach programme.

This presentation on Stone and Bronze Age axes complemented a display of neolithic flints with a hypertext information terminal. Visitors were asked to choose the right materials to make an axe, to place the handle in the best position, and suggest how to sharpen the blade. Some 70 screens of information took users 'from knowing nothing about prehistoric axes to knowing quite a lot', Buchanan said.

He estimated that the programme took about three weeks to prepare. The Education Department drew up a storyboard, prepared text on a word processor, and assembled images for scanning. Buchanan then put the package together in two days. He admitted that the displays 'probably looked home-made', but also noted that the curators thought visitors should be encouraged to look at the objects rather than a slick colour picture alone – the information systems are designed to enhance the displays, not replace them.

This is certainly a valid point in view of the many elegant visual databases now employed elsewhere.

It will be interesting to see if some of these do not simulate a tour of the gallery or exhibition so effectively that a few visitors will be tempted to dispense with the rest of the visit.

The quality of the scanned images varied considerably according to the quality of the source. (Illustrations from an archæological catalogue of 1892 'scanned superbly – much better than some of the modern stuff'.) With the equipment they could then afford, the Museum opted to produce monochrome images and use spot colour to pick out keywords in the text display. Buchanan later saw the Design Museum's *Study Collection Database* and observed that 'the quality of the images is dreadful, but they serve the purpose'.

There were only two hardware problems. One was occasioned by the design of the touchscreen, which was supplied as a cowling to fit over a standard PC: despite 'thinking long and hard' about the best height for the screens, the team did not anticipate that visitors – particularly the children – would *lean* on the cowling and, temporarily and inadvertently, incapacitate the touchscreen. (Field testing often reveals critical differences in attitude, expectations, abilities and behaviour between the production team and the target audience.)

Buchanan himself took responsibility for the other fault: in packing the hardware for delivery to the first outreach exhibition, he left most of the cable connections in place, thinking that this would make the system easier to move and install. One of these took a knock during delivery and incapacitated the keyboard (which, fortunately, is only used for inputting, and not in the display itself).

To save the technical staff the responsibility of starting up and shutting down the systems every day, both were designed so that the security staff had only to plug them in every morning, and unplug them at night.

Visitors had few problems with the systems, although Buchanan observed children in the Discovery Room explaining the use of a computer mouse to their parents and teachers. The mouse, he found, confused many new users until they made the connection between the mouse in the hand and the cursor on the screen: 'they tended to watch the mouse rather than the screen', he said.

Of two other computer-based exhibits in the Discovery Room at the time, Buchanan described one from British Gas that 'had a little man running about the house opening windows ... it was so complex that nobody could use it', and said another, from Atari, 'was very good but not reliable'. In keeping with its mission, there were no instructions posted in the Discovery Room; however, Buchanan felt that more encouragement for first-time users would have helped the *Wealth of a Nation* display.

GLASGOW'S GLASGOW

Cognitive Applications, whose work for the Design Museum and National Gallery is described on pages 46 and 56, produced an exhibition guide for *Glasgow's Glasgow* in 1990's European City of Culture.

The exhibition itself filled 22 vaults beneath Glasgow's Central Station, with 2300 objects from a Fabergé egg to football pennants. It included participative exhibits, 'experiences' (*Fly through Glasgow by computer*) and live entertainment. The interactive exhibition guide filled some 3000 screens with an indexed catalogue of all 2300 exhibits, together with cross-referenced background information and floor plans of the site. Visitors controlled the guide from a touchscreen with only four commands: GO ON, GO BACK, GO TO THE INDEX, and HELP.

The project was completed to a strict deadline within three months. There were neither resources nor time to create more than the minimal amount of data unique to the exhibition guide, so the display depended on material already produced for use in other parts of the exhibition.

Fortunately, this included captions for every object on display, which had been prepared on a Cardbox database. This proved a happy coincidence of both technology and style, for captions and text screens employ similar conventions to convey essential information succinctly, and it proved relatively easy to recycle information from the database in the exhibition guide.

With training from Cognitive Applications, the exhibition's senior researcher, Francis McKee – whose computer experience hitherto had amounted to word processing – assembled and edited 600 screens of illustrated text in less than a month. Although it was certainly convenient to have so much data already on a computer, the researcher's intimate knowledge of the subject matter contributed materially to the successful delivery of the finished product.

Given the tight deadline, Cognitive Applications elected to work with software which they already knew well, and the package was developed on an Apple Macintosh with HyperCard. However, it was destined to run on IBM-compatible portable computers donated to the show by Olivetti. Desktop Design (like Cognitive Applications, based in Brighton) helped to develop and transport the data from the Mac to the PC.

Glasgow's Glasgow opened on Friday, April 13th. The tight production schedule and unpropitious date notwithstanding, Alex Morrison of Cognitive Applications, writing in the August edition of *Museums Journals*, reported only two hardware faults, half-way through the show's six-month run: a touchscreen broken in transit, and one failed hard disk. He also noted some unexpected enthusiasm for the system: visitors asked if they could buy a copy on disk, and teachers used print-outs from the system during class visits to the exhibition and in related projects afterwards.

TATE GALLERY, LIVERPOOL

Art and technology met with conspicuous success in the Tate Gallery, Liverpool. *Sculpture Interactive* is a multimedia database which can change and grow in response to the needs and interests of its users.

The project was based on the work of Henry Moore and his contemporaries, in a collaboration between the Gallery and BBC Milton Keynes, with sponsorship from the Henry Moore Sculpture Trust, hardware donations from Apple UK and Pioneer Laser Technology, and support and services from the European Community's COMETT project, Microvitec, Bit 32 and Mersey Television.

The project was conceived as a case study. Moore became the obvious subject when an exhibition at the Royal Academy brought a large collection of his work under one roof (albeit briefly – the production team was given six nights to photograph there). Photos of individual sculptures, mostly shot in isolation, are complemented by images of work in locations from open fields to inner city precincts (complete, in some cases, with a backdrop of car parks or graffiti which is an observation in itself).

Selected works can be viewed from different angles and distances as though walking toward, away or around the work. Surrogate walks tour Moore's studio in Much Hadham and the studios of four other modern British sculptors.

The package uses Apple's familiar point-and-click routines with keyword menus, directional arrows and icons. The gallery's delivery system combines a computer screen for text and digitized stills with a video monitor in a purpose-built housing. The computer screens are uniformly arranged to provide background information for every item (for example, the title and location of each work, or the date and source of each radio interview or TV clip).

Some 200 individual sculptures are represented, in anything from 3 to 180 still images apiece. This collection is complemented by images of work in other media (including sketches, drawings, paintings, catalogues, books and posters), together with video clips of sculptors at work, explanations of technical processes such as bronze casting, records of major exhibitions such as *Moore in Venice* in 1976, and interviews with Moore and his contemporaries from TV and radio archives.

Verbatim subtitles on the computer screen allow users to follow the interviews word-for-word, and suggest a range of secondary uses in language teaching. A large collection of quotes is reproduced on text screens which users can access from a main menu that summarizes the key phrase from each quote. (A keyword search would be a useful enhancement of this feature.)

Some pieces were shot under different lighting conditions to illustrate the powerful effects of light and shade: a light bulb icon indicates when images can be studied in this way. Visually impaired users can switch between positive and negative images of the digitized still frames, and a variety of control systems (including a mouse, trackerball and bar-coded notebooks), together with proprietary products, allow greater access for disabled users who cannot manipulate conventional input devices.

Other versions developed during the pilot phase include a presentation system with a video projector, and a 'videotheque'. Even with both sides devoted largely to video, the LaserDisc holds some 5000 still frames and an extra hour of sound recordings on its second audio track.

The original design of the database favoured users with some previous knowledge of Moore's life and work. The main index to his sculpture, for example, was divided geographically, to reflect the locations of collections in Liverpool, Leeds, Much Hadham and elsewhere. This may reflect the physical limitations of the real world, but is not of primary importance within the infinitely flexible world of the hypermedia database. However, the package also includes an authoring facility which could be used to create presentations which examine the material in different ways. The MediaBase Resource in fact allows any user – art historian or schoolchild – to create a multimedia essay by 'collecting' still images, audio clips and moving footage from any part of the database, and arranging this material into a personal commentary or analysis.

Each item of audio-visual material can be accompanied by up to 1000 words of text and 90 seconds of spoken commentary digitally recorded, edited and stored entirely within the computer. The essays themselves become interactive – later readers can examine various parts of each essay independently. When the system was launched, the BBC's Senior Producer Martin Wright observed that, although intended to be accessible to art historians within three minutes of use, the package took most adult users 10 minutes or so to navigate successfully – although he estimated that the average schoolchild would master it within 90 seconds.

VICTORIA & ALBERT MUSEUM, LONDON

The T.T. Tsui Gallery of Chinese Art opened within the V&A in 1991. Four compact interactive terminals, in blond wood housings, stand in the two wide aisles of the gallery. The tops of the square terminals are slanted, to present a small screen set in a translucent dark green glass shield. Text printed there in English and Mandarin introduces the system:

THERE IS NO SOUND
THERE ARE FIVE SILENT FILMS; EACH LASTS TWO MINUTES.
YOU CAN INTERRUPT AT ANY POINT.

Around the screen, six circles with bilingual labels invite the user to TOUCH THIS BUTTON to start, or select a presentation. Sensors beneath the touch points control a LaserDisc player within the housing, which presents the appropriate segments of video. At the end of each presentation, the system reverts to an attractor screen, which displays a tiny window of moving footage from each segment, and then loops to a map of China.

In fact, some of the presentations last as long as four minutes, and a thorough tour can take up to a quarter of an hour, although the user can interrupt simply by pressing the button for another segment. The system is silent, using short bilingual captions to identify the subject or date of the video footage.

Monuments presents a brief survey of sights including the Great Wall, the Forbidden City and, as a rather surprising finale, the Bank of China building in Hong Kong. *A Quarter of the World: People 1500-1990* combines old photos and prints with recent footage, to juxtapose images of Court ladies with farm machinery, or Peking Opera with a modern wedding. *Living off the Land* uses film shot in rural China since the 1970s to illustrate scenes of agricultural labour and the country life of towns and villages. *Making Things,* similarly, combines modern scenes of basket-weaving and calligraphy with footage of jade carvers taken in Peking in the 1930s.

Art under the Microscope uses original footage created with equipment more often used for exploratory surgery than art history. Four artifacts – a porcelain dish, a bronze jug, a silk wedding curtain, and a sedan chair – are examined in graduated detail from a single overview to close-ups at actual size, and at powers of x20, x50, x100 and even x200, with text labels to identify each level of magnification. The highest power reveals the individual fibres in the intricately embroidered curtain, and minute patterns in the organic material of the dish and jug. Unfortunately, there is no key to the location of the objects in the gallery, nor is it immediately apparent (amidst so much archival footage) that these images are drawn from the V&A's own collection.

The protective shield on top of the unit bears most of the inevitable indignities: the systems enjoy virtually constant use during peak hours, and are usually smeared with fingerprints as a result. However, the white lettering on the shield was silk-screened on its underside, as a simple strategy to anticipate the effects of some thousands of touches.

The package was produced by Line TV, a London-based production company with extensive experience of interactive video. The systems were to be simple, elegant and relatively inexpensive (a report in *Museum Development* magazine suggests that hardware and software together amounted to around £10,000 per unit). They are controlled by a timer, initially set to the museum's official opening and closing times, but later adjusted to accommodate special events such as private views and extra openings during popular temporary exhibitions.

The principal design challenge was to find photographic material with the right content and technical quality, without sending a camera crew to China, or paying high fees to a film library. The Chinese embassy in London supplied a vault of material, which director Tymn Lintell spent many hours vetting, for both visual quality and content. It was particularly important to find short clips that could communicate immediately, without audio, to an audience from a very different culture.

The disc's producer, Piers Lea, observes that 'there seems to be a notion that, if you have masses of archival material, you can just bung it on a disc and make presentations from it. But there is real skill, hardly recognized, in finding the right material to tell a story, make it interesting and keep the audience entertained.'

American producer Ken Christie took a similar view in the January 1984 issue of *The Videodisc Monitor,* discussing the use of LaserVision at the EPCOT Center in Orlando, Florida. When American Express offered an archive of 50,000 colour slides, '[I thought] the production would be a piece of cake', he said, 'After all, we only needed 800 or so'. In the event, 50 slides had the content and quality required, and 'we paid several slide brokers for the other 750'.

WELLCOME INSTITUTE, LONDON

The Wellcome Institute for the History of Medicine will open an exhibition centre in its new building in London's Marylebone Road in 1992. Two production companies are developing radically different multimedia resources for the Wellcome Trust.

Media Projects International (MPI) is producing an interactive image base from material in the Trust's Iconographic Collection. This involves photographing over 50,000 images, in black-and-white and colour, at full size and in detail. The Collection ranges from prints and paintings to sculpture and even some archival film. Little of the source material in any archival collection will precisely fit the dimensions of a modern video screen, so a computer graphics system provides uniform backgrounds, and scales and stores images for transfer to disc.

The collection will be presented on a network of terminals which combine two screens under the central control of Wellcome's mainframe computer. This holds a large database, which can be accessed at different levels of detail. Images from LaserDisc will be displayed on a high-resolution monitor, with associated text from the database on a separate computer screen below.

The system will be used by visitors to the exhibition, and specialists including students, curators, scholars and researchers for film, television and publishing. Like the Jersey Museum's archival project, the interactive image base will provide this wide audience with a valuable service, and protect the collection itself from handling: colour slides or prints will be supplied for reference on request.

Extensive research by MPI included many practical tests to determine how to photograph a large and diverse range of items to minimize risk to the objects and maximize production. This led to a number of fundamental decisions on both technical and administrative questions. (The proposal for the project was in fact written in two versions: one in everyday English, and one in technical terms, 'to prove that we knew what we were doing' says MPI's Malcolm Lewis.)

On the technical side, processing was refined to improve the quality of reproduction on disc: saturated colours and gilding, for example, both tend to glow too vividly on a video screen. Lewis observes that 'taking pictures properly in the first place is a great help' when photography is only the first stage of the production process, and believes that careful planning can effect considerable economies of scale if one photo session can produce colour and black-and-white shots which satisfy the various needs of curatorial staff and education, publishing and marketing services. High-quality images, he says, are a hedge against technological innovation, if they can later be processed for use in new display and archiving systems.

With some 5000 rolls of film to shoot, time and motion studies contributed to the administrative schedule. The single shot which is executed in a split-second takes some minutes to set up, and getting the objects and the camera to the same place can be time-consuming. (MPI briefly considered adding a night shift, but the project has fortunately kept to schedule.) MPI, like many others, believe that it is usually better to turn directly to experts than attempt to adapt in-house resources to work for which they were not designed. As Lewis observes, internal services are usually busy enough without taking on large special projects.

In working with the database, Lewis observes that 'the slowest part of the project is data entry'. Audio-visual expertise aside, he feels that key personnel must include 'qualified researchers who know what they're looking at and can get information out of curators' *and* extremely fast and accurate keyboard operators to enter the data.

(This is a view reflected by Philip Phillips of the Liverpool Museum, who also undertook a multimedia cataloguing project. At a Museums Association seminar on Computers in Museums in 1988, he emphasized that data entry is a time-consuming task which demands strict control over both terminology and computer coding.)

Lewis also emphasizes the importance of assigning a single identification number to each image, which will follow it through all subsequent processing, both curatorial and technical. The Iconographic Collection will take two-and-a-half years to capture as an interactive database; it is scheduled for completion in 1993.

In a separate project, New Media are producing a series of interactive displays and educational games for the new exhibition galleries, on subjects as diverse as *The Population Explosion, Protein Synthesis* and *Gene Therapy*. The games make extensive use of high-quality computer graphics, for delivery on standard PCs with touchscreens – a strategy which allows for rapid response times, and the introduction of new games and information in the future.

EUROPEAN MUSEUMS & GALLERIES

The use of multimedia and interactive systems is developing rapidly across Europe, for gallery guides and information packages, and archival image bases. This section briefly notes some of the more interesting and innovative projects.

IMAGE BASES

The Centre Nationale d'Art Plastique in Paris produced *Video Museum*, a database of 10,000 images, as a pilot programme for a project which could eventually record 140,000 images of 20th century French art from 80 museums – some 80% of the nation's public collections in this field.

The Complete Collection of the Mauritshuis Museum uses a remote control keypad to retrieve images and text which describe the collections of The Hague's principal fine arts museum. The still images can be run as a short 'slide show' with music.

The Courtauld Institute in London used microfiche to catalogue photographs of 17th century Dutch painting from the Witt Library, supported by a grant from the Getty Foundation. The Courtauld and the V&A have also worked with Emmett Publishing on resources discs for images of fashion, art and architecture. The original choice fell on LaserDisc, but publisher Jim Emmett says that, with the emergence of new CD formats, he elected to watch the market before making a final choice of media.

The Danish Royal Library in Copenhagen is working with the National Museum and DRIVE (the interactive video unit of Danmarks Radio) on a study which could ultimately record some ten million photographs on an optical disc database for use by visitors, curators and scholars.

The Getty Foundation has sponsored a project to put some 25,000 prints from the Warburg Institute in London, and the Biblioteca Herziana in Rome, onto videodisc, with associated textual references.

The Maritiem Museum Prinz Hendrik in Rotterdam combines a LaserDisc resource base, of some 27,000 images, with an Apple Macintosh HyperCard control program (developed by the Dutch interactive media centre, NIAM) for use in its galleries and library.

The Musée Albert Kahn in Boulogne uses videodisc to present 8000 images from film, autochromes and stereographs taken by Albert Kahn's photographers between 1909 and 1931.

The Musée d'Orsay in Paris has released its LaserDisc of 11,000 images of 2,100 works, and a guided video tour of the collection; Voyager produced complementary software (at $124.95 for the disc, $99.95 for the software).

The National Gallery of Sweden led a consortium of Swedish museums in the production of an interactive image base of some 10,000 images of Swedish art from 1500 to 1940. The National Gallery will provide half the material, with another 20 galleries contributing about 200 pictures apiece (both full images and details). The project was conceived as a learning exercise to develop expertise within the participating institutions and produce resource material of value to them all.

The Russian Museum in St. Petersburg is transferring reproductions of the 340,000 items in its collection from 35mm film to LaserDisc, in a project with Interactive Learning Productions (ILP) of Newcastle. Additional material, providing details of selected works, is being shot on 5"x4" inch plates by Ken Morse of London. ILP developed a retrieval system for the collection which searches on criteria such as artist, date, subject, media or material, using a trackerball to select items from a menu. The discs will be used in Russia for education and tourism, and sold internationally as resource material for other institutions; ILP report that, with a list price of around £500 per disc (and another £1800 for the delivery system), they received 500 orders soon after the first disc was launched.

The Union Française des Arts du Costume in Paris put 48,000 images of 30,000 items of 20th century *haute couture* on a videodisc, with control software which allows users to store small digitized images from the disc on a computer's hard disk for use in scholarly documents and studies.

The Vasari Project is a European consortium funded by the European Commission's ESPRIT initiative to investigate the use of digital techniques to record and manipulate high-resolution images for visual databases.

INTERPRETATIVE EXHIBITS

Battle Abbey, by the site of the Battle of Hastings in southern England, uses IV in a presentation on *The Monastic Life*, produced for English Heritage by 3T Productions. The menu of the programme is presented as a plan of the mediæval Abbey: visitors touch any part of the site to see illustrations and re-enactments of daily life there (with soberly costumed actors and plenty of period detail).

The Granada Television Studio Tour in Manchester features an interactive video quiz on TV trivia. *Telequiz* has five levels of difficulty, with daily, monthly and yearly prizes for top scorers. Questions are chosen at random from a large database, and both speed and accuracy are needed to win. The package was also produced by 3T Productions.

The Guinness World of Records uses a Sports Databank (produced by Xebec) and LaserDisc for AV at exhibits in London, Hollywood, Las Vegas, Niagara Falls, New York and Singapore.

The Hayward Gallery in London used 3D computer graphics from LaserDisc in an exhibition in 1988 of Leonardo da Vinci's architectural drawings, studies in geometry and use of perspective. The package was developed by IBM's Scientific Centre in Winchester. In 1992, IBM developed 3D reconstructions of objects and sites for a temporary exhibition on Pompeii at London's Accademia Italiana.

The Heureka Science Centre in Helsinki complemented a programme on nuclear energy with a quiz which begins with a simulated journey from the sun into the earth's core and the heart of an atom. Two questions determine whether players should attempt 'trainee' or 'champion' levels. Copies of the videodisc were given to other Finnish science centres.

Ipswich Eastern Farm Park is now the permanent home of *The Food Chain*, an information programme on British food and farming originally developed by 3T Productions for a nationwide tour in 1989. The package illustrates the industry required to feed the nation, using the example of a simple meal in a fast food restaurant, and provides insights into each user's own eating habits.

The Kunsthistorische Museum of Vienna produced *Vienna Interactive*, a videodisc-based guide to the art and history of the Austrian capital. A similar package, on the Old City of Stockholm (*Gamla Stan*), was produced principally for use in schools by TelePedagogic of Nynäshamn, using HyperCard software by MultiMedia of Stockholm.

Tussauds' London Planetarium adapted published videodiscs and live satellite images to create a range of short, silent video clips in the Space Trail, a holding area which keeps visitors entertained before scheduled performances of the laser Star Show in an adjacent auditorium. Events organisers also bring their own software for use on the systems when the galleries are hired for private functions.

The Museum of the Moving Image (MOMI) in London's South Bank Centre supports its myriad AV exhibits from a bank of 72 LaserDisc players in a central control room. Displays include material transferred from archival footage in many film and video formats and participative exhibits such as newsreading and an 'interview' with film critic Barry Norman. (Although the Museum uses actors in character as guides to various parts of the exhibition, the interactive displays are not formally supervised. Thus it is observable that, while some visitors are confident about role-playing in front of friends and strangers, the participative exhibits are not always used as their designers might have wished.)

The National Maritime Museum in Greenwich, near London, uses two information systems – one interactive, one passive – to help visitors find facilities such as shops and restaurants, and to promote special events and temporary exhibitions; both packages were produced by Xebec Multi Media Solutions. Other interactive systems include a computer-based package on the steam paddler *Reliant*.

The National Museum of Denmark produced a two-screen multimedia system for the Freedom Museum in Copenhagen to mark the 50th anniversary of the German occupation of the country. The database contains some moving footage and 11,400 still images (photographs, documents and artifacts), with commentary in Danish and English. The package is sold as resource material for education.

The Natural History Centre at Liverpool Museum uses *British Garden Birds*, one of a few educational discs published by the BBC when LaserVision was first launched. *The Geology Information Programme* is used in a learning centre which offers visitors a view of specimens from the reserve collection.

Genoa's Natural History Museum featured a wholly computer-based display, developed by Hay Space of Milan, in a quincentenary exhibition which examined the exchange of flora and fauna between Europe and the Americas since 1492.

The Oslo Museum of Science and Industry won the Museums Association prize for Most Innovative Product in 1987 for *Science at Your Fingertips*, produced by ScanoVision and Futuremedia. The touch-screen programme presents information on eight topics in Swedish and English.

The Ulster Museum in Belfast commissioned an eclectic pilot videodisc (with segments on historical photographs, marine invertebrates, artifacts from the Armada, and modern art) from the Northern Ireland Centre for Learning Resources.

The Vasa Museum in Stockholm renovated its displays in 1990 to include IV and hypermedia systems which help visitors to appreciate what the warship might have been like before it capsized, on its maiden voyage, in Stockholm harbour in 1628. The ship was excavated in the late 1950s, and is well preserved, but computer graphics, slides and video help to provide visitors with detailed views of the ship, a surrogate walk on its decks, and a simulation of a 17th century warship under sail.

The Wimbledon Lawn Tennis Museum in south London uses LaserDisc and multimedia in five linear and three interactive programmes. Archival footage presents a history of champions from the 1920s onward, while *The Players' Databank* provides up-to-the-minute coverage in text and photos, which the museum's curators can update after every tournament. A trivia quiz employs conventions from the game of tennis, and awards successful contenders with a trophy. The multimedia presentations were produced by Xebec Multi Media Solutions.

Windsor Safari Park, near London, provides interpretative material about the animals in residence there through an information programme developed by 3T Productions. The package combines LaserDisc and computer graphics to illustrate *Body Design, Conservation, Food and Feeding, Habitat, Lifestyle* and *Facts and Figures* about each animal.

Work in progress in January 1992 included a project developed by New Media for the museum of a London borough, which combines many features separately explored in other installations described here. The collection will encourage visitors to investigate everyday objects from the recent past to make connections between the archival material and their own daily lives, and to add anecdotal material to the database. Through interactive technology, a local history collection will become a community project that grows with the participation of its visitors.

CONSORTIA AND CO-PRODUCTIONS

Co-productions, between national partners and across national borders, undoubtedly represent the way forward for many undertakings in the 1990s.

The Dante Project, coordinated by the CTU (University Television Centre) of Milan University, combines the text of *The Divine Comedy* with resources including commentaries, translations, a dictionary and encyclopædia of similes, and 500 digitized images from 14th and 15th century miniatures.

The Parthenon is the first of a projected series of discs coordinated by the Etablissement Public du Grand Louvre with the British Museum, the Bibliothèque Nationale, the Acropolis, Athens Museum, Skelpturhalle (Bale) and the Louvre's own department of Greek, Etruscan and Roman antiquities.

The Sacred Way will combine photographic images of today's ruins and reconstructions in 3D computer graphics to provide an information system for visitors to the site of Eleusis, an ancient religious centre near Athens. The graphic reconstructions support a surrogate walk through the site as it might have appeared 2500 years ago. The project is being funded by the European Commission to demonstrate the potential of multimedia systems to professionals working in archæology, museums and the 'heritage industry'. It is being produced by New Media, working with the Benaki Museum, Athens, the Institute of Archæology, Oxford University, Mechanical Intelligence UK and MNEME.

VIDEODISCS IN MUSEUMS 1988

The Multimedia and Videodisc Monitor, the leading newsletter in the US market, first published *Videodiscs in Museums* in January 1988. The second edition was published in January 1992; it would, therefore, be churlish to steal its thunder (*address, page 81*). However, some findings from the first report are of interest, particularly *because* they represent a market still struggling with some of the very first videodisc players and interactive systems.

The report discussed 96 applications, the majority in the US. Seventy participated in a brief survey, and all said that they 'would choose videodisc again'. They registered their other views on a sliding scale from 0 (extremely dissatisfied) to 5 (thoroughly satisfied):

- Interactive videodisc solved our particular problem effectively *Mean response 4.79 out of 5.0*
- Videodisc has proven cost-effective *4.56*
- Other museum personnel have responded favourably to the system *4.53*
- We have found the system dependable *4.51*

The report's one negative response came from The Carter Presidential Museum and Library in Atlanta, Georgia, which had suffered problems with overheating and power failure (faults rarely experienced in the current generation of solid-state systems).

The most extreme demands were made by the US National Park Service, with 337 systems nationwide. Some used the venerable Pioneer PR-7820 (the Model T Ford of the LaserVision) although most had newer Sony players, presenting linear video or interactive programmes run from the player itself.

Many of these stand virtually out-of-doors, with minimal protection from the elements – one actually sits on a glacial moraine. Even so, the Park Service reported an average downtime of less than 15%. The discs were produced with existing footage wherever possible. Several include a choice of up to five languages, and all are close-captioned. (The US Parks Service is still commissioning new programmes for its IV network.)

A separate feature in the *Monitor* of January 1984 reported on the first year's experience at the EPCOT Center in Orlando, Florida. The Center had been using 272 LaserDisc players to present 83 programmes (30 interactive, 53 linear) on 63 videodiscs. All 272 were left running 24 hours a day, although the precision motors which control the laser positioning were turned off separately for six to eight hours overnight. With these early players, EPCOT experienced some problems, but the *Monitor* calculated a mean time between failures of some 4600 hours.

Videodisc producer Ken Christie observed that visitors to EPCOT spent from 2 to 13 minutes with the interactive systems (although seating in some areas may have encouraged longer visits). As the initial fascination of the touchscreen wore off, he noted that visitors 'can become restless if they have to wait through unwanted material'.

SOME AMERICAN INSTALLATIONS

Videodisc and interactive multimedia systems are used in American museums, galleries and private exhibitions from the Smithsonian Institution to the Texas Book Depository. A fuller list of institutions and their projects begins on page 85. Some of the more interesting projects are described briefly here.

THE SMITHSONIAN INSTITUTION

The Smithsonian Institution was one of the first to use IV, and to exploit new multimedia publishing opportunities. Some $2.7m was spent on the information centre in the Great Hall of its 'Castle', which opened in 1989 with 13 interactive information terminals in cherrywood kiosks. Four are dedicated to *Capital Attractions*, which describes 120 museums, galleries, historical sites and other places of interest in the Washington area, with information on opening hours, transport and so forth. The rest describe the 14 museums within the Institution itself, and the National Zoological Park, with information on the location, services and special events of each. The package won the 1990 'Nebbie' (from the Nebraska Videodisc Design/Production Group) for 'Best Consumer Achievement'.

The units run on a Local Area Network (LAN), so information can be centrally updated and distributed. The systems are complemented by two 'orientation theatres', electronic wall maps, scale models of the Institution, and rolling screens of daily events. All the interactive systems offer a choice of seven languages (English, Spanish, French, German, Chinese, Japanese and Arabic), although first-hand reports indicate that not all languages are supported equally. The Information Centre expects about three million visits a year.

Treasures from the Smithsonian is one of the first consumer titles in the American CD-I market, and the Institution also produces a series of double-sided LaserDiscs (distributed by Lumivision for $29.95 each) on the history and contents of its collections. Titles include *Dinosaurs, First Ladies, Flying Machines, Gems and Minerals, Insects,* and *The National Zoo.* A three-hour presentation on two discs, *John Fitzgerald Kennedy: A Celebration of His Life and Times*, sells for $59.95.

US LIBRARY OF CONGRESS

The US Library of Congress launched the six-year American Memory Project in 1989 to disseminate fragile and uncatalogued material (including manuscripts, photographs and films) on LaserDisc and CD-ROM. The Library's curators identified material for cataloguing and distribution to other libraries, and some $900,000 in first-year government funding will be devoted to cataloguing alone.

The Legislative Governance of America will include 1500 daguerreotypes of 19th century legislators and public figures, *Visions of America: Seeing the Nation through the Lens of Popular Art and Culture* will include 20 hours of ethnic folk music and 3000 Currier and Ives prints, and *American Places and People* will bring together many local history books and documents which are now out of print. By September 1991, the Library had pilot-tested the first of the material, and was looking for 30 diverse sites nationwide to initiate an evaluation project.

VIRTUAL REALITY

The Mandala Experience is a 'virtual environment' designed by Vivid Effects, which uses a video camera to record the movements of the person in front of the screen, and project this image so that the user appears to enter the picture and interact with objects there – playing instruments and bursting bubbles to release flights of doves, all computer-generated.

Connecticut Museum of Natural History features a display employing Virtual Reality (VR), the 3D technique used in *The Mandala Experience* to create the illusion of putting the viewer literally into the picture. VR is most popular in games, but its familiar apparatus – helmets which concentrate the player's vision on special optical effects, and DataGloves which simulate action within the virtual environment – are often seen in television programmes which, paradoxically, emulate a three-dimensional interactive experience in a flat, linear medium. (The BBC's Education Service simulated VR techniques in *The History File*, which presents artifacts and documents from the Roman world with state-of-the-art production techniques.)

OTHER PROJECTS

The Amparo Foundation of Mexico chose CD-I for a projected 30 information terminals while its new museum in Puebla was still under construction. The collection covers Mexican history from the pre-Columbian period to the Spanish conquest. Visitors use headphones and touchscreen terminals in each room to learn more about displays there. A timeline relates artifacts to contemporary events in other cultures, and the system includes games and other activities in Spanish, English, French and Japanese.

The Asian Art Museum in San Francisco used LaserDisc and an Apple Macintosh in a collaboration with Xerox Palo Alto Research Center (PARC), to produce the *Electronic Sketch Book of Tibetan Thangka Painting*. The package explores techniques, elements, samples and curatorial analyses of an ancient and endangered tradition.

The Avery Architectural and Fine Arts Library at Columbia University will convert 45,000 architectural drawings to videodisc, with 6500 text records for cross-referencing. The Research Libraries Information Network (RLIN) will be used to create and store the database, and display the drawings and accompanying text on separate screens.

The Boston Computer Museum bucked the trend for miniaturization with a working model of a personal computer two storeys high. The screen alone measures 10 square metres/108 square feet, and the keyboard is 38m/125' wide. Six interactive learning stations, jointly developed by OWL International and VideoLogic, illustrate the development of computer hardware from silicon chips through to workstations, and explain how software is made.

Cleveland Public Library chose Panasonic's write-once Optical Memory Disc Recorder (OMDR) to catalogue a collection of 42,000 movie stills. Work began in January 1990 and the Library launched the first volume of 12,000 images (from movie titles A-D in the alphabetical collection) that November. The PC-based retrieval system allows keyword searches by title, year, studio, genre, actors or scene. Standard 5"x4" prints are sold for $2, and larger prints are available on request.

Ford's Theater in Washington DC now houses a collection devoted to the president who was assassinated there in 1865. The Lincoln Iconography exhibit includes a digital image base of 1300 stills of portraits and memorabilia, originally shot on 35mm film and digitally encoded for storage on a write-once Recordable LaserVision (RLV) disc. The exhibition includes a matched set of portraits of Lincoln from youth to maturity, scaled to the same size and position on the screen. The eyes remain fixed as the portraits slowly dissolve from one to the next to create the impression of the president's face visibly ageing on the screen.

The National Portrait Gallery in Washington DC transferred 3000 images to CD-ROM as part of a pilot study on the use of compact disc imaging systems in museums and galleries.

St. Louis Zoo opened a $17.8m high tech education centre called *The Living World* in 1989. The exhibits include 30 computer stations, 32 large and 20 smaller video screens, 10 LaserDisc players, 85 short films, a $500,000 donation of Macintosh computers from Apple, and a robot that looks like Charles Darwin. A 12-minute show in the Ecology Hall simulates a space flight around the earth with satellite photos from high definition videodisc. The Zoo and Voyager produced an IV package on animals, *The World Alive*, which sells for $49.95.

The San Diego Museum of Man commissioned a new interactive display from archival footage for *The Human Life Cycle*. Interactive Arts produced new voice-overs, music and graphics to adapt material on genetics, reproduction, fertility, pre-birth and birth. The source material includes still photos and microcinematography of cells and embryos.

Sega Enterprises, a leading maker of computer games, used live action video and special mirrors to create the illusion of holograms in the *Hologram Time Traveller*. The arcade game must be played in the dark, and has a smaller angle of vision than a true holographic display, but Sega say that it can be produced, with video sources including LaserDisc and CD-I, more cheaply than holograms. *RePlay* magazine's survey of earning performance ranked the game as the third most popular in the US in 1991.

The Tech 2000 Center in Washington DC's Techworld Plaza is a showcase for interactive LaserDisc and multimedia applications of all kinds, from training and education to public exhibitions and consumer products. It includes exhibits from the Chicago Sports Museum and Hall of Fame, the George Eastman Collection, and the National Gallery of Art, as well as *The Image Bank*, a videodisc-based resource for commercial photo libraries.

The United Nations began to transfer 175,000 photos from its collections (including those from UN agencies such as UNICEF, UNESCO and WHO) to LaserDisc in the spring of 1989. The whole collection will probably fill four disc sides, and the UN plans to distribute the collection commercially.

The Interactive Video Science Consortium is supported by some 20 museums of science and technology. Another seven American museums formed the Museums Education Consortium (MEC) to develop interactive media for use in schools, colleges and homes. Their first project, a resource disc on Impressionist and post-Impressionist art from 147 collections worldwide, had its debut at New York's Museum of Modern Art (MoMA). Split-screen presentations allow users to compare landscapes with film footage of the same scenes as they now appear.

Some of the conceptual problems observed in curators' first encounters with new media were described by Californian artist Michael Pye in *The Independent on Sunday* (11 February 1990). He designed a computer graphics city which was displayed, on LaserDisc, at MoMA. 'The museum never did grasp why there was no photographer's credit', he said, 'or how a picture could be printed from a digital data tape'.

DOMESDAY AND AFTER

The Domesday Project was initiated by the BBC to commemorate the 900th anniversary of the Normans' epic survey of their new kingdom with an interactive multimedia database which would record British life in the 1980s. IV was seen as the platform on which to develop an information system which would carry schools, libraries and commercial users forward into the 20th century. It didn't quite work out like that, but *Domesday* has attained the peculiar status of a project whose failure inspired a second generation to try again, and do it better.

Domesday combined statistical data, maps and other conventional resources on the National disc with a Community disc that drew on material contributed by some 14,000 schools, groups and clubs throughout the UK. Some guidelines ensured that data about the communities were collected uniformly across the country; however, the contributors could also write essays, and submit up to three photographs, on any topic they thought interesting, important or typical of daily life in the mid-1980s.

The result was a fascinating archive of the predictable, the weird and the wonderful. The discs are already a time capsule – not least because the project also developed a unique videodisc format, LV-ROM, for which no more than a handful of other titles were developed subsequently.

Domesday thus provided an interesting model for information providers, and an object lesson for hardware developers. (The decisions to avoid in future included hardware choices which disdained *de facto* standards, software which immutably recorded statistical and other data which were soon outdated, and prohibitively high prices for both software and hardware which were not justified by comparison to products already available in other media.)

Nonetheless, in the wake of *Domesday*, similar projects sprang up in other parts of the world. Some, like Canada's Jean Talon Project, appear to have foundered; others, like Denmark's *SIULLEQ*, are already published.

500 Años después was conceived, like *Domesday*, to commemorate a momentous anniversary: 1992, the year which sees the quincentenary of Columbus' voyage from Barcelona to the Americas, as well as the Olympic Games in Barcelona and a World Fair in Seville. Nine partners from Spain, Portugal, Italy and Denmark produced the multimedia resources (80% of which were drawn from existing archival material) for transfer to a dynamic editorial database in Madrid. The collection concentrates on Italy, Spain and Portugal, Latin and Central America. The project has funding worth over 6m ECU, a quarter of which comes from the European Commission.

SIULLEQ (Greenlandic for *The First*) was undertaken by DRIVE, the interactive unit of Danmarks Radio, with the Autonomous Government of Greenland, to produce a multimedia database on the political, cultural and natural history of Greenland, for use in schools, libraries and museums. The package, in Danish and Greenlandic, combines LaserDisc and CD-ROM for use with an Apple Macintosh, and is now published on disc for 4000DKr.

Nordic Life Today was a separate three-year project led by the Educational Software Group of the Nordic Council of Ministers. Funding of 6.5mDKr was drawn about equally from the five participating countries, with a top-up from the Nordic Council.

Heritage: New Zealand was sub-titled 'a national community project for 1990', and followed *Domesday's* approach in organizing hundreds of schools and community groups to gather data and pictures to create a multimedia 'snapshot' of life in New Zealand in 1990. Unlike *Domesday*, the project also aimed to develop a low-cost platform employing industry-standard hardware.

Numbers 2000 was conceived to produce a multimedia database of Jewish history and culture, with the specific aim of making Jewish children in many countries more aware of their heritage. The package was designed to include data on specific places (accessed through maps) and surrogate walks through sites of particular significance, but also includes games and 'typical histories' which follow ordinary individuals through daily life and historic events. This is the only privately-funded project of its kind so far: a non-profit trust administered donations from institutions and charities in Israel, America and Europe.

HDTV IN MUSEUMS & GALLERIES

A number of projects now under development propose to use HDTV systems in public displays, and proponents of the technology are producing demonstration and broadcast material in the new format.

JAPAN

The most ambitious use of HDTV is that made by NHK, the Japanese Broadcast Corporation, at its Hi-Vision Gallery in the Museum of Fine Arts, Gifu. Some 700 of the museum's 2000 works (mainly by symbolist painters) are stored on HD videodisc, for display on three systems. A 42-seat theatre uses a rear projector with 2.8m/110" screen; two variations offer 1.5m/60" and 74cm/29" screens. Each picture can be retrieved by artist's name, date or genre, and an accompanying database provides information about artists and work in the collection.

The Hi-vision Library in the Sezom Museum of Art in Tokyo offers a similar facility. High quality images illustrating the historical background of artists and works in the collection were created with the Hi-Vision Videograph (HVD-1100) system for display on a 2.3m/90" rear projection screen. The museum's offices use a HVD-1100 retrieval system with 50cm/20" Hi-Vision display and scanner.

The first of Pioneer's CUBE Hi-Vision videodisc players were used in the Flower and Green Expo '90 exhibition in Tokyo, which also employed CD-I among its high tech information services.

CHRIS PARSONS PRODUCTIONS

Chris Parsons Productions of Bristol has actively promoted HDTV for some years, principally through plans for an 'electronic zoo' such as already exists in Tokyo. Filming an actual bird colony or wildlife park with three cameras simultaneously produces the footage to simulate a panoramic view like that seen from the windows of a real observatory or hide.

Managing Director Chris Parsons says that a prototype of the observatory was so well received at a meeting of some 300 guests of the Natural History Museum Development Trust, that some visitors were 'almost hypnotized' by the demonstration, and reluctant to leave. (The demo was subsequently adjusted to 'close the blinds' on the observatory windows at the end of every show.) Sony helped to underwrite some of the development and filming for Parsons' pilot project.

With such a new platform, Parsons has faced the unenviable task of creating awareness while keeping abreast of enhancements to the technology itself. He notes that early work had to be shot on high-quality film, but the introduction of portable high-definition video equipment should soon allow developers to shoot directly on high-definition stock.

Parsons believes that HDTV's 16:9 aspect ratio will foster new shooting techniques, and that high-definition makes new demands on make-up artists and set and costume designers. 'On HDTV a painted tile floor looks like a painted floor rather than the real thing', he says.

The Bristol Development Corporation has included plans for an HD observatory among the high tech exhibits in its proposed new Environmental Centre, and Chris Parsons is also talking to institutions in the US and Europe, including one which is considering a wholly electronic zoo, with no animals at all.

OTHER HDV & HDTV APPLICATIONS

In 1989, the St. Louis Zoo was the first institution in the US to use high definition video for a public display. A Sony HDL-2000 videodisc player presents a 12-minute show which includes animated satellite photography to simulate a flight around the earth.

The World of Coca Cola pavilion in Atlanta, Georgia, installed America's first permanent HDTV cinema in 1990, with a Sony HDIH-2000 5m/200" front-screen projector and HDL-2000 videodisc player.

The National Gallery of Art in Washington DC received a similar system from Sony in 1991 for use in an 80-seat cinema. Gallery Director J. Carter Brown said, 'The use of high-definition TV and videodiscs will enable the Gallery to project, without deterioration, images of detail and colour that would be impossible to replicate on videotape'.

Corporate customers for Sony's HDTV hardware have included Ford UK (for Ford Germany), who used HDTV in a high-end CAD/CAM (computer-aided design/manufacture) development. Fiat in Italy and BMW in Germany have used HDTV to produce commercials and promotional material.

The Tower of David Museum in Jerusalem has used two HDTV programmes, on a Sony HDL-2000 disc player: a 10-minute cartoon made by an Italian producer, and a scenic tour of Jerusalem. Both can be displayed on an ordinary Barco projector.

Computer graphics companies such as Symbolics and Quantel in the UK are examining how HDTV images can be used with computer 'paint' packages.

PROFESSIONAL ADVICE

Some common themes emerged in discussions with the developers whose work is described here, which potential clients or producers may wish to consider.

PARTNERSHIP

The publishing career of New Media's Dick Fletcher began with books and now extends to media invented within the last decade. He advises all newcomers to 'begin in a small, modest way now and upgrade. There is a path through the new technologies, and it can be self-financing.' He advises museums to treat funding as an integral part of any project, and exploit new publishing opportunities, but says 'The key to all this is not the finished product but the relationship. You can hire the most creative person imaginable and end up with a lousy product because the relationship wasn't right.'

Fletcher says 'you have got to get to a point where somebody trusts somebody', and believes that clients who encourage sharp competition for short contracts will not achieve good working relationships with their suppliers. In working with museums, 'We start by agreeing timescales, rules and roles, but we allow curators to run the project if they wish'. Fletcher sees the developer's role as enhancing, with special expertise, information provided by the client: 'We always work with specialists. We know nothing. We try to be a blank page.'

Malcolm Lewis of Media Projects International emphasizes 'the need for both parties to understand each other's culture, hierarchy and sense of time', and stresses that the supplier must liaise directly with '*one* person at the museum invested with sufficient authority to make decisions'.

He also speaks of the need for strong project management within the museum to combat 'curatorial disease', an academic and political condition which impedes both research and production. (Another contributor, who worked at a national gallery before taking a university post, says that she is admired in her new job for her skill 'in dealing with individuals who are regarded as "difficult" by people who have never worked in a museum'.)

The bedrock, Lewis says, must be strategic planning on the largest possible scale – 'even if you only do a little to start with'.

FUNDING

New Media's Dick Fletcher recalls that his company's involvement with *The Living Arctic* began with a report which was used to help attract sponsors. He sees this as a vital part of the programme-maker's role. 'We have been involved in raising funds in this way for many projects', he says. 'We like it.'

New Media's client in that project, Dale Thompson of the Canadian Secretary of State's Department, recognizes that 'the public sector is more and more facing the need to operate jointly or with sponsors – BROUGHT TO YOU BY IBM CANADA'. His advice is simple: 'Get as many partners as you can up front: the ability to sell a package, especially in this market, is very dependent on people's feeling that they had some input. There are more thought-provoking questions than technology, and if you have the skill and capacity to address these, not only can you reduce costs and get a little economy of scale going, but you are more competitive.'

Malcolm Lewis of MPI advises all museums and galleries to 'consider the revenue-earning potential of your project – which implicitly includes a valuable PR gain'.

PRODUCTION VALUES

The importance of video is a volatile issue in contemporary multimedia design, where so many critical trade-offs surrender high visual quality for high storage capacity. The interesting paradox is that audiences rarely appreciate the technical constraints which inform so many creative decisions, but neither do they demand that everything on screen perform like a blockbusting film or a top-grossing arcade game. Interaction with simple graphics can be more engaging than inaction during video clips. Audiences are usually much less critical of production values than are, say, other programme-makers.

Neil Sandford, erstwhile MD of The Soft Option, believes that 'The trick is not to attempt broadcast television's production values unless you're prepared to spend that much time and money to make one hour of programming. Good graphics are better value than bad video – TV uses 10 different shots of the same thing just for variety.'

He cites an analysis by Ged Fitzgibbon of Computer-Aided Training Services, which tracks the relationship between development time and hours of use in different industry sectors. Where an hour of good computer-based training can be produced within 75 hours, broadcast television typically requires 1500 to 2000 hours of work for every hour of programming. This can rise to 4000 hours and more in advertising (a single advert will of course be broadcast many times, but a successful training package may equally be used for years without modification, at a much lower cost).

Sandford also argues that a successful prototype is a cost-effective foundation for a larger project – and one which identifies serious problems at the outset, when they can be most effectively addressed, can be particularly cost-effective. Prototyping can demonstrate how production costs can be reduced through the creative use of existing resources (not only video and still images, but information and intellectual property as well).

Sandford also believes that successful programme-makers must pay serious attention to the implementation and evaluation of their work: 'You must evaluate the first disc before you start the second', he says, and cites examples of large corporate users who commissioned too many IV packages too quickly, and failed to learn or profit from experience as their courseware libraries grew.

He compares the production process to a form of mountaineering, with a peak dividing the creative struggle uphill and the gradual journey downhill through the more predictable stages of production and implementation.

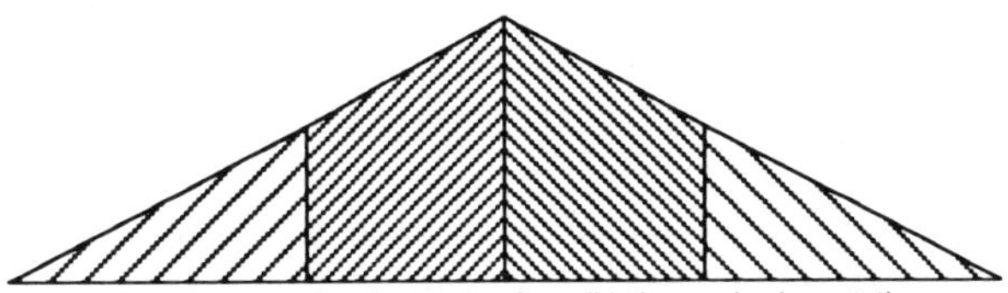

SANDFORD'S TRIANGLE

'At the first sign-off point, between analysis and creative development, you're still going uphill and things will still change. If you lock yourself into a production schedule or budget at that point, and start to go downhill from there ... you *really* will go downhill from there. The key to success is to ensure that nothing changes after you reach the end of the creative phase.'

'The danger with an IV budget', he says, 'is that the budget goes 40% for analysis, 40% for development, 40% for consolidation ... by which time you're 20% over budget and you have nothing left for implementation and evaluation.'

A good budget, he maintains, 'should be divided equally between the four stages from analysis to implementation – with no more than 5% variation either way'. The whole production cycle, he maintains, should be a constant process of analysis, design, development, implementation and analysis, so that something is learnt from each stage and from each project before the next begins.

Museum visitors are often, by definition, predisposed to discover new ideas and experiences. Mick Landmann of Vivid Interactive has helped to develop interactive systems for environments where they are neither expected nor familiar. *Renewable Energy*, as well as its permanent display at the Manchester Museum of Science and Technology, is used in touring exhibitions by the Department of Energy. An earlier project, *Family Credit*, was developed for the Department of Social Services.

Like *Renewable Energy*, the package combines digital audio, digitized colour photos, computer-generated graphics and text on an Apricot Xen-i computer. Four simple questions determine users' family and financial status, to advise whether and how to apply for Family Credit. A dot matrix printer gives each user a summary of the data which they supplied – take-home pay, number and ages of children, and savings – which determine eligibility for the benefit.

The system was first field tested with a week apiece in a Gravesend shopping centre, a Woolwich leisure centre and the Lewisham branch of the British Home Stores. Passers-by responded well to the system, but the designers felt that the prototype contained too much information.

'It became confusing to some users', Landmann says, 'and the help function, although common in training packages, only added to the consumers' uncertainty'. It was dropped, the choices on any one screen were restricted to three, and control keys on the front of the unit were reduced to four. The system underwent a second round of field testing, and was conspicuously successful in attracting the target audience.

(The system was on display at the National Interactive Video Centre in London, with no more signage or instructions that it would normally have displayed at a shopping centre. This writer observed a team of carpet cleaners one Saturday morning, literally starting the inert system by plugging it in, following the instructions on the screen to get it running, and taking turns to use it to learn if they would be eligible for Family Credit – entirely without prompting, or indeed any previous experience of interactive information systems.)

Mick Landmann observes: 'There is a tendency when one is designing any project to include features which seem attractive to the designer but which can prove to be totally inappropriate for the user. We forget that the world out there is full of people who have never been exposed to this sort of thing in any shape or form. And simple *can* be good. Something that may seem simplistic to developers may be a totally new experience to its audience. We need to be designing for *them* – not for other programme-makers or awards committees. We should definitely involve more potential users in the development process.'

SOFTWARE SALES

It is notoriously difficult to supply hard marketing data in the multimedia industry. However, some anecdotal evidence has emerged:

In April 1989, the *Monitor* reported that the US National Gallery of Art had sold 3000 copies of its $95 LaserDisc, which combines a video tour of the gallery with still frames from the collection. The Gallery's Ruth Perlin attributed a rise in sales to the development of a HyperCard stack (that is, associated indexing and interpretative programs for an Apple Macintosh) by Voyager.

Dr. David Clark claimed that, with sales of 300 discs in six years (1983-1989), the University of London Audio Visual Centre's *Programme Development Resource Disc* (£63) was Britain's best-selling LaserVision disc.

The Dutch Royal Library in The Hague spent Fl.120,000 in 1987/88 to produce a LaserDisc of 4000 images from books of the 9th to 16th centuries, and 2800 woodcuts from Netherlandish incunabula. The disc sold for Fl.1100 with a printed index of frame numbers, or for Fl.2100 with IBM-based retrieval software. Its distributors, MEMO Media Marketing of Utrecht, reported that over 100 copies were sold in the first two years, to libraries and museums.

The Academy of Natural Sciences in Philadelphia had sold seven copies of *Discovering Dinosaurs* when the *Monitor* conducted a survey in 1987. Former Director of Exhibits Laurence Hutchinson felt that IV contributed to a 'tremendous increase in attendance [which] more than justified the cost of the exhibit' and said that the sale of the package to other museums 'helped us to recover some of our costs'.

Even now, a LaserDisc on a general subject may sell in the low thousands, where one on a specialist topic would sell in the low hundreds. The great hope, of course, is that the new CD platforms, or even the re-launch of LaserDisc to consumer markets, will create the hardware base on which software sales necessarily depend.

SAMPLE PRICES

Prices for IV and CD software vary wildly, according to factors including the quantity of data on the disc, the media mix and the cost of producing it, and the 'felt fair' price in the target market (generally, low for education and consumer titles, high for professional training). Corporate IV prices range from £8250 for a single financial training programme to £400 for a two-part health and safety package.

Animals in Motion* and *Women in Motion
(On-Line Entertainment)
CDTV: £29.95 each
Two discs of archival photographs which can be captured for use in computing or desktop publishing.

The Bible and Gustave Doré's Engravings
(Fototeca Storica Nazionale)
CD-ROM: 390,000 Lira
The Bible, with 520 engravings, plus an analytical index with over 4000 references.

Illustrated Shakespeare
(Animated Pixels)
CD-I: $50/£34
The Works, with 1000 illustrations.

Images of the French Revolution
(Pergamon Press and Bibliothèque Nationale, Paris)
LaserDisc: £250 to £43,000
Some 35,000 images from the revolutionary period. A printed catalogue and PC database are available separately or with the disc. The £43,000 version includes 995,000 pages of primary documents on microfiche.

The Louvre
(Laser Communications Plus)
LaserDisc: £100/$99.95 each
Three discs cover *Paintings and Drawings, Sculpture and Objets d'Art* and *Greek, Roman, Egyptian and Near East Antiquities;* Voyager has produced HyperCard stacks for use with the discs on an Apple Macintosh.

North Polar Expedition
(Virgin Games)
CDTV: £49; LaserDisc: £99/$99
Adventure game based on Sir Ranulph Fiennes' 1979/82 Transglobe Expedition. The IV package includes a double-sided LaserDisc, two PC disks, manuals, maps and a teacher's guide.

EXCHANGE RATES (APRIL 1992)

This table provides a quick key to prices and costs quoted here in various national currencies.

	£	$		£	$
Austria	20.00	11.56	Netherlands	3.20	1.84
Belgium	58.47	33.79	Norway	11.15	6.44
Canada	2.06	1.19	Portugal	244.42	141.28
Denmark	11.02	6.36	Spain	180.42	104.29
France	9.62	5.56	Sweden	10.32	5.96
Germany	2.84	1.64	UK	1.00	0.58
Ireland	1.07	0.62	USA	1.73	1.00
Italy	2142.50	1238.43	ECU	1.39	0.80

NETWORKS & USER GROUPS

There are now a number of international organizations dedicated to creating and maintaining links between museums, galleries and related companies and services in the multimedia market.

Three special interest groups are cited in the first part of this report: the European CD-I Association (page 28), European DVI Developers' Group (page 24) and European LaserDisc Association (page 26).

BRITISH INTERACTIVE MULTIMEDIA ASSOCIATION (BIMA)
6 Washingley Road
Folksworth, Peterborough PE7 3SY (UK)
Contact: Jane Callaghan
Tel. +44(0)733 242370, Fax +44(0)733 240020

The British Interactive Multimedia Association (BIMA) is the representative body for the interactive multimedia industry. BIMA has five main objectives: (1) to represent the multimedia industry in its relationships with customer groups and government; (2) to promote a wider understanding of the benefits offered by interactive multimedia; (3) to establish professional standards for the multimedia industry and to communicate these as widely as possible; (4) to provide a forum for the exchange of views among members; (5) to develop and maintain the high standards of professional practice. The association was established in 1985 as the British Interactive Video Association (BIVA); membership is open to any company with a demonstrable interest in multimedia and its applications.

ECSITE
The Nuffield Foundation
28 Bedford Square, London WC1B 3EG (UK)

The European Collaborative for Science, Industry and Technology Exhibitions (ECSITE) is an organization for science centres and museums, exhibition designers and others 'with a commitment to hands-on exhibition techniques'.

EUROPEAN MUSEUM NETWORK
Museon, Stafhouderslaan 41
2517 HV s'Gravenhage (Netherlands)
Contact: Friso Visser
Tel. +31(0)70 338 1405, Fax +31(0)70 350 6319

An EC-sponsored RACE project with eight museums and four technical partners. Each museum contributes information on 100 objects for an information system accessible to all.

EVAC & EVAIN
The Library, Suffolk College, Rope Walk
Ipswich 1P4 1LT (UK)
Tel. +44(0)473 211214, Fax. +44(0)473 230054
Contact: Jeremy Rees

The European Visual Arts Centre (EVAC) in Ipswich is scheduled to open about the same time as the new British Library. The European Visual Arts Information Network (EVAIN) provides an information service for galleries, museums, libraries and others with an interest in multimedia displays and image bases. EVAC maintains the Image Technology in European Museums and Art Galleries (ITEM) database on behalf of EVAIN.

INTERACTIVE DESIGNERS ASSOCIATION (IDA)
PO Box 1414, London SW5 9UY (UK)

The IDA's monthly meetings help interactive designers to keep abreast of developments in the industry and to meet each other. Membership is open to anyone involved in the development of interactive or multimedia programmes.

INTERACTIVE MULTIMEDIA ASSOCIATION (IMA)
Suite 240 – 800 K. Street, N.W.
Washington DC 20001 (USA)
Contact: Ede Bartell
Tel. +1 (202) 408 1000, Fax +1 (202) 408 0361

The Interactive Multimedia Association (IMA) is an international trade association for corporations, institutions and individuals who produce equipment, software and programming for multimedia applications in education, training, entertainment and information management.

MUSEUM DOCUMENTATION ASSOCIATION
347 Cherry Hinton Road
Cambridge CB1 4DH (UK)
Tel. +44(0)223 242848

MUSEUMS ASSOCIATION
34 Bloomsbury Way
London WC1A 2SF (UK)
Tel. +44(0)71-404 4767

Two central sources of information on museums, installations, networks and developments of all kinds, with information services including publications and events.

PUBLICATIONS & EVENTS

DIRECTORIES

DVI Technology Applications Developers Directory 1991, Digital Vision 1991

European Multimedia Yearbook 1992, I M Publishing 1991

International Directory of Interactive Multimedia Producers, Multimedia Computing Corporation 1990/91

The CD-ROM Directory 1992, TFPL Publishing 1991

The CD-ROM Market Place 1991, Meckler 1991

The Multimedia Directory, Philips Publishing 1990

The Videodisc Compendium, Emerging Technology Consultants 1991

INTERACTIVE MULTIMEDIA

AMBRON, S. & HOOPER, K., *Interactive Multimedia*, Microsoft Press 1990

BERRY, Dianne & HART, Anne (ed.), *Expert Systems: Human Issues*, Kogan Page 1989

BOVE, Tony and RHODES, Cheryl, *Macintosh Multimedia Handbook*, QUE Corporation 1990

BROWN, P.J., *HyperDictionary*, Kogan Page 1988

FELDMAN, Tony, *Multimedia in the 1990's*, The British National Bibliography Research Fund 1991

KWEE, Nick, *Lenk's Laser Handbook: Featuring CD, CDV and CD-ROM Technology*, McGraw Hill 1991

LUTHER, Arch C., *Digital Video in the PC Environment*, McGraw-Hill 1989

McINTOSH, Stephen Ian, *The Multimedia Producer's Legal Survival Guide*, Multimedia Computing Corporation 1990

MEGARRY, Jacquetta, *Compact Discs and Computers: Converging Technologies*, Kogan Page 1989

RICHARDSON, A., *Exploiting Digital Communications*, National Computing Centre 1988

TUCKER, Richard N. (ed.), *Interactive Media: The Human Issues*, Kogan Page 1989

CD-I

HOFFOS, Signe, *The CD-I Designer's Guide*, McGraw-Hill 1992

LAMBERT, S. & SALLIS, J. (ed.), *CD-I and Interactive Videodisc Technology*, Howard W. Sams 1986

CD-ROM

ARMSTRONG, C.J., and LARGE, J.A., *CD-ROM Information Products, Evaluation Guide and Directory*, Gower 1990

ARMSTRONG, C.J., and LARGE, J.A., *CD-ROM Information Products, Volume 2: The Evaluative Guide*, Gower 1990

BOWERS, Richard, *Selected Resources for CD-ROM and New Media Publishers*, Information Arts 1991

FELDMAN, Tony, *CD-ROM (Publisher's Guide Series)*, Blueprint Publishing, London, 1987

LAMBERT, S. & ROPIEQUET, S. (ed.), *CD-ROM: The New Papyrus*, Microsoft Press 1986

SHERMAN, Chris (ed.), *The CD-ROM Handbook*, Intertext and McGraw-Hill, 1988

INTERACTIVE VIDEO & VIDEODISC

BINDER, Roberta, *Videodiscs in Museums*, Future Systems 1991

PALMER, R., *Management Guide to CBT/Interactive Video*, National Computing Centre 1988

PALMER, R., *What is CBT/Interactive Video?*, National Computing Centre 1988

PERLMUTTER, Martin, *Producer's Guide to Interactive Videodiscs*, Knowledge Industry Publications 1991

SOUTER, Gerald, *The Disconnection: Interactive Video and Optical Disc Media*, Knowledge Industry Publications 1988

DESIGN AND PRODUCTION

ADDIS, T.R., *Designing Knowledge-Based Systems*, Kogan Page 1986

ANDERSON, Carol, and VELJKOV, Mark, *Creating Interactive Multimedia: A Practical Guide*, Scott Foresman Macintosh Computer Books 1990

ARWADY, Joseph, and GAYESKI, Diane, *Using Video: Interactive and Linear Designs*, Educational Technology Publications 1989

BARRETT, Edward, *Text, Context and Hypertext: Writing with and for the Computer*, MIT Press 1988

BERGMAN, Robert, and MOORE, Thomas, *Managing Interactive Video/Multimedia Projects*, Educational Technology Publications 1990

BERK, Emily, and DEVLIN, Joseph, *The Hypertext Hypermedia Handbook*, McGraw-Hill 1991

BUNZEL, Mark J. and MORRIS, Sandra K., *Multimedia Applications Development: Using DVI Technology*, McGraw Hill 1992

IMKE, Steven, *Interactive Video Management and Production*, Educational Technology 1991

IUPPA, Nicholas, *A Practical Guide to Interactive Video Design*, Croom Helm 1984

IUPPA, Nicholas, and ANDERSON, Karl, *Advanced Interactive Video Design*, Knowledge Industry Publications 1988

JOHNSON, Kerry A & FOA, Lin J., *Instructional Design: New Alternatives for Effective Education and Training*, NUCEA/ACE & Macmillan (USA)

LAUREL, Brenda (ed.), *The Art of Human-Computer Interface Design*, Addison-Wesley 1990

LAURILLARD, Diana, *Interactive Media: Working Methods and Practical Applications*, Ellis Horwood 1987

PALMER, R., *Designing and Using CBT/Interactive Video*, National Computing Centre 1988

PEARSON, Don (ed.), *Image Processing*, McGraw Hill 1991

SCHULTZ, JOHN AND BARBARA, *Picture Research: A Practical Guide*, Van Nostrand Reinhold 1991

SWAIN, Dwight and Joyce, *Scripting for the New AV Technologies*, Focal Press 1991

WOODHEAD, Nigel, *Hypertext and Hypermedia*, Addison Wesley 1991

FURTHER INFORMATION

PUBLICATIONS & EVENTS

MAGAZINES, NEWSLETTERS & JOURNALS

CD-ROM International
Transtex International, 11 Rue du Marché Saint-Honoré
75001 Paris (France)

Digital Vision
Suite 202, The Foundry, 156 Blackfriars Road
London SE1 8EN (UK)

Electronic & Optical Publishing Review
Learned Information, Woodside, Hinksey Hill
Oxford OX1 5AU (UK)

European Multimedia Bulletin
Suite 202, The Foundry, 156 Blackfriars Road
London SE1 8EN (UK)

Information Media and Technology
Hatfield Polytechnic, P O Box 109
Hatfield AL10 9AB (UK)

Inside IT
Tossa House, Main Road
Smalley, Derby DE7 6EF (UK)

Interactive Multimedia
Sigma Press, 1 South Oak Lane
Wilmslow, Cheshire SK9 6AR (UK)

Interactive Media International
I M Publications, 104 St. John Street
London EC1M 4EH (UK)

Intermedia
International Institute of Communications
Tavistock House South, Tavistock Square
London WC1H 9LF (UK)

Laser Disc Review
Philip Parr, P O Box 526
London SW10 9AB (UK)

Multi Media
High Text-Verlag, Gabelsbergerstraße 52
D-8000 München 2 (Germany)

Multi-Media Computing
Nanton Press, Postbus 93, 3720 AB Bilthoven
Soestdijkseweg 332N
3723 Bilthoven (Netherlands)

Multimedia and Videodisc Monitor
Future Systems Inc., P O Box 26
Falls Church VA 22046 (USA)

Multimedia Review and *HD World Review*
Meckler Ltd, 247-249 Vauxhall Bridge Road
London SW1V 1HQ

Optical Information Systems
Meckler, Grosvenor Gardens House
London SW1W 0BS (UK)5

Screen Digest
37 Gower Street
London WC1E 6HH (UK)

ANNUAL CONFERENCES & EVENTS

CD-ROM Expo (USA)
IDG World Expo
+1 (508) 879 6700

The Digital Image: Pictorial Arts and Information Technology (Scandinavia)
Norwegian Computing Centre
+47(0)2 46 69 30

Electronic Imaging and the Visual Arts (UK)
Brameur
+44(0)252 313437

European Conference on Multimedia and CD-ROM
Microsoft
+49(0)211 55 62 81

European Multimedia Computing Show
Nanton Press
+31(0)30 290 644

Eusidic Annual Conference (UK)
Eusidic
+44(0)249 814584

Image Processing (UK)
Blenheim Online
+44(0)81 868 4466

Interactive Learning (UK)
John Barker
+44(0)332 881779

International Conference on Multimedia and CD-ROM (USA)
Microsoft
+1 (203) 352 8224

New Information Media (UK)
The British Library
+44(0)71 323 7041

Multimedia (UK)
Blenheim Online
+44(0)81 868 4466

Online Information (UK)
Learned Information
+44(0)865 730275

TIME Europe and CD-ROM Europe (UK)
Lowndes Exhibition Organizers
+44(0)71 383 3323

INTERACTIVE MULTIMEDIA EXHIBITS

The lists below summarize museums, galleries and private exhibitions in Britain, Europe, Canada and the US which are known to have used multimedia systems or to have projects in hand. Some simply use LaserVision for the delivery of sound and pictures; some developed systems for use only in temporary exhibitions. The list is drawn from many primary and secondary sources, and updated regularly; however, travellers to distant parts are advised to confirm that the systems which they wish to see are still on display before embarking.

UK

Avon	**Avon Forest Park** Forest guide and interpretative material
Beamish	**North of England Open Air Museum** *Building*
Belfast	**Ulster Folk and Transport Museum** Various gallery displays
Belfast	**Ulster Museum** 'Videodisc sampler' of gallery displays
Birmingham	**Birmingham Museum, Gallery 33** *The Collectors in the South Pacific*
Bracknell	**Bracknell Museum** Video and audio-only from LaserDisc in various displays
Bradford	**Cartwright Hall** Information on exhibitions and events at the Hall and other attractions in Bradford
Bradford	**National Museum of Photography, Film and Television** Archival film and television footage and participative displays
Cambridge	**Museum of Archæology and Anthropology** *The Nagas*
Canterbury	**The Canterbury Tales** Walk-though experience with scenes of life in Chaucer's England
Dover	**White Cliffs Experience** Interactive LaserDisc in *Historium*, other linear video from LaserDisc
Edinburgh	**Royal Museum of Scotland** Hypertext gallery displays for temporary exhibitions and Discovery Room
Glasgow	**Glasgow's Glasgow** Exhibition guide to *European City of Culture 1990*
Gloucester	**National Waterways Museum** *Canal Builder, Operating a Lock* and *Steering a Boat*
Gloucestershire	**Crickley Hill Country Park** Random access videotape on topics including geology, archæology and wildlife
Greenwich	**National Maritime Museum** Gallery guides, educational computer display about the steam paddler *Reliant*
Halifax	**Children's Museum** *Safety*
Hampshire	**New Forest Museum and Visitors' Centre** Flora and fauna of the New Forest
Hastings	**Battle Abbey Visitors' Centre** *The Monastic Life*
Hawkshead	**Hill Top Collection** The life and works of Beatrix Potter
Ipswich	**Eastern Farm Park** *The Food Chain*
Jersey	**Jersey Museum** *The Story of Jersey*, image base of photographs
Liverpool	**Beatles Museum** Various displays using interactive LaserDisc
Liverpool	**Liverpool Museum** Resource Centre, BBC's *British Garden Birds* in the Natural History Centre
Liverpool	**Tate Gallery Liverpool** *Sculpture Interactive*
London	**Bank of England Museum** *The Bank Today*

London	**British Library** Various intractive displays (opening 1993) plus CD-ROM for schools
London	**British Museum** *The Making of England* temporary exhibition (1992) and CD-ROM for schools
London	**Courtauld Institute** Image base of the Witt Library on microfiche
London	**Design Museum** *Design a Toothbrush, Study Collection Database*
London	**Geological Museum** Exhibits include *Treasures of the Earth* and *The Story of the Earth*
London	**Guinness World of Records** *The Sports Databank*, plus linear video from LaserDisc
London	**Hayward Gallery** 1988 exhibition with 3D computer graphics based on Leonardo da Vinci's architectural drawings
London	**Imperial War Museum** Databases, archival footage and 'experiences'
London	**Museum of Mankind (British Museum)** *The Living Arctic*
London	**Museum of the Moving Image** Linear video on LaserDisc, interactive displays include simulated TV interviews
London	**National Gallery** Micro Gallery
London	**Natural History Museum** *Eyewitness Testimony, Super Memory, The Carbon Cycle, Mammals* and *Insects, Dinosaurs*, many wholly computer-based interactive systems
London	**Planetarium/Madame Tussauds** Space Trail (images from LaserDisc and live satellite links), laser Star Show, videowall
London	**Science Museum** Gallery displays on topics including cells, molecules, industrial chemistry, and outer space
London	**Victoria & Albert Museum** *Images of China's Past and Present*
London	**Warburg Institute** Funding from the Getty Foundation to produce a visual database of 25,000 images
London	**Wellcome Institute** Various gallery displays, Iconographic Collection (opening 1992)
London	**Wimbledon Lawn Tennis Museum** *So you think you know about Wimbledon?*
Manchester	**Granada Television Studios (Backstage Tour)** *Telequiz*
Manchester	**Manchester Museum of Science and Technology** Exhibits include *Good Planet Guide, Journey to the Planets, Renewable Energy*
Northumberland	**Kielder Visitors' Centre** Simulation of water transfers in the Borders
Norwich	**Castle Mall** Information terminals (opening 1993)
Norwich	**Tourist Information Centre** Guided walks around *Norwich: A Fine City*
Oxford	**The Oxford Story** Dark ride through multimedia 'experience' of university life in history
Sheffield	**Ponds Forge** 1991 World Student Games – The Museum of Human Achievement and Adventure
St. Andrews	**British Golf Museum** Various gallery exhibits, *Great British Golf*
Tunbridge Wells	**A Day at the Wells** Scenes of life in a Georgian spa town enhanced by new technology
Windsor	**Windsor Safari Park** Exhibit on ecology and conservation in Windsor Safari Park
Worcester	**Battle of Worcester Museum** Costumed displays about the English Civil War
Yelverton	**Buckland Abbey** History of Buckland Abbey
York	**Automata Museum** Displays on the history of automated machines
York	**Jorvik Centre** Britain's first and most famous ride through historical reconstructions
York	**National Railway Museum** Gallery displays

EUROPE

Amsterdam	**Royal Library** LaserVision resource disc with some 7000 images of mediæval miniatures and woodcuts
Barcelona	**Caixa de Pensions' Science Museum** Various gallery exhibitions
Boulogne	**Musée Albert Kahn** *Collections Albert Kahn*
Copenhagen	**Freedom Museum** *The German Occupation of Denmark, 1940*
Copenhagen	**Museum of Zoology** Display on Danish bird life based on the BBC videodisc *British Garden Birds*
Copenhagen	**National Museum** Gallery exhibits in the ethnography department
Copenhagen	**Royal Library** Development work on interactive visual databases
Florence	**Uffizi Project** 'Videotel' information using slow-scan video to transmit data
Florence	**Various sites** 150,000 images of art treasures in Florentine museums and churches
Frankfurt	**German Film Museum** Clips from historic films on LaserDisc
Frankfurt	**Jewish Museum** Videodisc database of Jewish culture, tradition and history
Frankfurt	**Rosso Bianco Collection** *Cars and Their Engines* from the world's largest collection of sports and racing cars
Genoa	**G. Doria Natural History Museum** 1492-1992 commemoration on the exchange of flora and fauna between the Old and New Worlds
The Hague	**Mauritshuis** LaserVision resource disc of images from the collection
Helsinki	**Finnish Science Centre** *Evomedia* and *The Nuclear Quiz*
Italy	**Magic Bus** *Animali della Savana*
Madrid	**Centro de Arto de Reina y Sofia** 7000 works of 19th and 20th century art on LaserDisc
Madrid	**El Museo Nacional Centro de Arte Reina Sofia** 7000 images of Spanish and Latin American art on videodisc
Mannheim	**Museum of Technology and Work** 12 LaserDiscs and 45 systems with database of technology exhibits
Munich	**Mercedes Benz Museum** The history of the Mercedes
Münster	**Westphalian Museum of Natural History** Various displays using interactive LaserDisc
Netherlands	**Rijksdienst vor de Beeldonde Kunst** Worldwide travelling exhibition on fine art from the Netherlands
Oslo	**Museum of Science and Industry** *Science at Your Fingertips*
Paris	**Centre Georges Pompidou** Various temporary exhibits
Paris	**Centre Nationale d'Art Plastique** *Video Museum* network project
Paris	**EuroDisney** Dark rides, various AV presentations supported by multimedia
Paris	**Louvre** LaserVision resource discs
Paris	**Musée Carnavalet** *Mozart à Paris* (temporary exhibition 1991/2)
Paris	**Musée d'Orsay** Information on artists and works in the collection
Paris	**Musée de la Villette** Various gallery exhibits and videowall
Paris	**Muséum d'Histoire naturelle** *On a marché sur la terre* (temporary exhibition 1991/2)
Paris	**National Army Museum** Videodisc database of 'military images'
Paris	**Union Française des Arts de la Costume** Videodisc database of 33,000 images of work from 20th century French couturiers

Rome	**Spada Gallery** A guided or self-paced preparatory tour of the Gallery, visual database of the collection
Rotterdam	**Anthropological Museum** Various exhibits with displays triggered by infra-red sensors
Rotterdam	**Maritiem Museum Prinz Hendrik** LaserVision resource disc of images from the collection, with HyperCard program
St. Petersburg	**Russian Museum** LaserDisc image base of 340,000 items from the collection
Spain	**Sociedad Estatal** Travelling exhibition on Spanish seafaring and the European discovery of America
Stockholm	**National Gallery** Image base of 10,000 items from the national collection and 20 other galleries
Stockholm	**Vasa Museum** Surrogate walk, simulated sailing and detailed inspection of the 17th century warship
Vienna	**Kunsthistorische Museum** *Vienna Interactive*
Consortia	***The Parthenon*** Videodisc-based resource bank coordinated by the 'Grand Louvre' project ***The Sacred Way*** Visitors' guide to Eleusis, with 3D reconstructions of the ancient Greek religious centre **Vasari Project** ESPRIT initiative to develop high-definition visual display systems and related software

USA

Nationwide	**US National Parks Service** 337 systems in 205 sites
US & Canada	**Guinness World of Records** LaserDisc and interactive displays in Los Angeles, Las Vegas, New York and Toronto
Albuquerque NM	**New Mexico Museum of Natural History** Two gallery displays on geology
Atlanta GA	**Carter Presidential Museum and Library** Nine videodisc programmes
Atlanta GA	**World of Coca Cola Pavilion** Various high tech displays including a permanent HDTV cinema
Baltimore MD	**Maryland Science Center** Various exhibits including *The Energy Glossary* and *The Whole TV Tour*
Boston MA	**Boston Computer Museum** *Building Computers*
Boston MA	**Museum of Science** Various exhibits on topics from *Cell Biology* to alcohol abuse
Boston MA	**Old South Meeting House** Interactive audio, including speeches, sermons and a prayer meeting on the night of the Boston Tea Party
Boston MA	**The Children's Museum** Adapted version of the *The First National Kidisc*, and two exhibits 'to dispel racial stereotypes'
Cape Canaveral FL	**Kennedy Space Center** Information units in Visitors' Center
Chicago IL	**Adler Planetarium** *Quest for Infinity*
Chicago IL	**Chicago Sports Museum and Hall of Fame** Databases on sports' stars in the Hall of Fame
Chicago IL	**John Shedd Aquarium** *The River Otter*
Chicago IL	**Museum of Science and Industry** Over 20 exhibits, including several on the exploration and exploitation of outer space
Cleveland OH	**Cleveland Children's Museum** *Over and Under Bridges*
Cleveland OH	**Cleveland Public Library** Videodisc-based archive of 42,000 movie stills
Cody WY	**Winchester Arms Museum** *Lock, Stock and Barrel*
Columbus OH	**Center of Science and Industry** *Mission to Mars*, simulated voyage to Mars and travelling multimedia exhibition
Cooperstown NY	**Baseball Hall of Fame** IV database on the game and its players
Dallas TX	**Texas Book Depository** Exhibit on the assassination of President Kennedy

Denver CO	**Museum of Natural History** Various IV exhibits on health care in the 'Life is Choice' permanent exhibition
Detroit MI	**Detroit Science Centre** Various permanent and temporary exhibitions
Endicott NY	**IBM Corporation Headquarters and elsewhere** Public information programme on IBM and its role in the local community
Far Hills NJ	**United States Golf Association Museum** Information, advice and quizzes on golf
Florida	**Tourist Information Centres on the Florida Turnpike** TouchGuide information services, using DVI technology
Fort Worth TX	**The Cattleman's Museum** Ranch life in Texas at the turn of the century
Greenbelt MD	**Goddard Space Flight Center** Satellite repair and the commercial uses of outer space
Indianapolis IN	**The Children's Museum** *Passport to the World* (toys and folk objects)
Jackson MS	**Mississippi State Historical Museum** *The Struggle for Equal Rights*
Kansas City MO	**Hallmark Visitor Center** Hallmark's accomplishments as greeting card manufacturer and corporate patron of the arts
Lexington KY	**American Saddle Horse Museum** Saddlebred World Championship Shows since 1917
Los Angeles CA	**California Museum of Science and Industry** Exhibits include *The Interactive Diner*, where a waitress comments on the nutritional value of your meal
Los Angeles CA	**Los Angeles Zoo** Adventure Island children's centre
Los Angeles CA	**Natural History Museum** Exhibits on various topics (Chaparral in the ecosystem, precious stones, dinosaurs)
Louisville KY	**Kentucky Derby Museum** *The Derby Time Machine* and *Place Your Bets*
Madison WI	**Wisconsin Historical Museum** *People of the Woodlands*, *Canoe Making* and *Old Land, New Land*
Milwaukee WI	**Milwaukee Public Museum** Atmospheric phenomena, flight and dinosaurs
Milwaukee WI	**Zoological Gardens** Two programmes on nutrition
Murray KY	**National Scouting Museum** Information on the Boy Scouts of America
Mystic CT	**Seaport Museum** Videodisc database of photos of famous yachts
New York NY	**American Museum of Natural History** Hall of Human Biology and Evolution (opening autumn 1992)
New York NY	**AT&T InfoQuest Center** Some 40 exhibits including holograms, robotics and voice recognition systems
New York NY	**Metropolitan Museum of Art** The Primitive Art Information System
New York NY	**Museum of Modern Art** MEC pilot disc of 1200 Impressionist and post-Impressionist works
Norwalk CT	**Maritime Centre** Various exhibits on maritime and local history, marine science, ecology and navigation
Oak Brook IL	**Ray Kroc Museum** Dioramas and interactive 'interviews' with the founder of the McDonald's fast food chain
Oklahoma OK	**Enterprise Square USA** 75 players in various exhibits on the principles of the free enterprise system
Orlando FL	**EPCOT** Some 80 programmes on science, space, ecology, selected countries and cultures
Philadelphia PA	**Academy of Natural Sciences** *Discovering Dinosaurs*
Philadelphia PA	**The Franklin Institute** Railroad exhibit (short films, still photos, games) and *The Fragile Earth*
Philadelphia PA	**Historical Society of Philadelphia** *Finding Philadelphia's Past*
Pittsburgh PA	**Carnegie Natural History Museum & Carnegie Science Center** Exhibits include an interactive planetarium and *Stratavator*, a simulated descent into the earth
Rochester NY	**George Eastman House** Museum guide and history of photography display
Sacramento CA	**Sacramento History Center** Various exhibits on local history and culture

San Diego CA	**San Diego Museum of Man** *The Human Life Cycle*
San Francisco CA	**Asian Art Museum** *Electronic Sketch Book of Tibetan Thangka Painting*
Santa Fe NM	**Museum of International Folk Art** Interpretative guide to objects on display in a 19th century 'period room'
Santa Monica CA	**J. Paul Getty Museum** Exhibits on topics including Greek vases and illuminated manuscripts
Saratoga Springs NY	**American National Museum of Racing and Hall of Fame** Exhibits include *Blueprint for a Champion* and Hall of Fame database
Seattle WA	**Woodland Park Zoological Gardens** Interpretative centre for the Northern Trail
Springfield MA	**Naismith Basketball Hall of Fame** Nine exhibits and games on basketball and its players
St. Louis MO	**Bowling Hall of Fame** Three exhibits including ' interviews' with bowling celebrities and a nine-screen videowall
St. Louis MO	**St. Louis Science Center** An adapted version of *The Earthquake Disc* (a published resource disc)
St. Louis MO	**St. Louis Science Zoo** *The Living World*
Tampa FL	**Museum of Science and Industry** Many exhibits on aspects of communication, language and telecommunications
Virginia Beach	**Virginia Marine Science Museum** *Open Ocean Dive*, a simulated trip 600 feet below sea level
Washington DC	**Ford's Theatre** *Lincoln Iconography*
Washington DC	**Library of Congress** American Memory Project
Washington DC	**National Archives** Short films on aspects of American history
Washington DC	**National Gallery of Art** Videodisc database of works from the collection, and a video tour of the Gallery
Washington DC	**National Geographic Society** Interactive theatre with multiple rearscreen projection and sixteen-screen videowall
Washington DC	**Navy Memorial and Visitors' Center** Biographical database with photos, on the Navy Memorial Log
Washington DC	**Smithsonian Institution** Visitors' guides and gallery exhibits
Washington DC	**Smithsonian Institution: National Air and Space Museum** Various exhibits on aviation and space exploration
Washington DC	**Smithsonian Institution: National Demonstration Laboratory** Demonstrations of interactive technology, technical assistance to museums and education
Washington DC	**Smithsonian Institution: National Museum of American History** Exhibits include *Information Age*, and the American Constitution
Washington DC	**Smithsonian Institution: Travelling Exhibition Services** Programmes from various parts of the Institution, adapted for use in travelling exhibitions
Washington DC	**TECH 2000** Over 90 interactive multimedia systems for training, education and entertainment
Washington DC	**United States Information Agency** *Information USA* and *Let's Speak English* (developed for a two-year tour of the USSR)
Wichita KS	**Lake Afton Public Observatory** Various programmes on astronomy
Williamsport PA	**Little League Baseball Museum** Little League World Series Highlights

CANADA

Banff, Alta.	**Banff National Park** Simulated golf game on this famous course
Calgary, Alta.	**Olympic Hall of Fame** Various gallery displays
Don Mills, Ont.	**Ontario Science Center** Examples of computers at use in daily life
Ottawa, Ont.	**Canadian Museum of Civilization** Galleries displays including *The Living Arctic* and *500 Images of Colonial Life in Upper Canada*
Ottawa, Ont.	**National Museum of Science and Technology** Videodisc supplies 246 radio and television news clips in one gallery display
Toronto, Ont.	**Royal Ontario Museum** *Halley's Comet*

ADDRESSES

This list represents museums and suppliers who contributed interviews and/or background information to this report. Organizations such as EVAIN, the Museum Documentation Association and Museums Association (*addresses, page 79*) can provide more information about other institutions; *The European Multimedia Yearbook* (*inside back cover*) lists suppliers and products in the multimedia market.

Keith Downes
3T Productions
50 Stamford Street
Ashton-under-Lyne, Lancs. OL6 6QH (UK)
+44(0)61 344 0908

John Keyworth
Bank of England
Museum & Historical Research Section
Threadneedle Street
London EC2R 8AH (UK)
+44(0)71-601 4444

Jane Peirson-Jones
Birmingham Museum
Dept. of Ethnography
Chamberlain Square
Birmingham B3 3DH (UK)
+44(0)21 235 4201

Peter Lewis
British Golf Museum
Bruce Embankment
St. Andrews, Fife KY16 9AB (UK)
+44(0)334 78880

Rowena Loverance
British Museum Education Service
Great Russell Street
London WC1B 3DG (UK)
+44(0)71 323 8511

Robin Sewell
Cambridge Multimedia Systems
St. Andrews
North Street
Burwell, Cambridge CB5 0BB (UK)
+44(0)638 743121

Jane Whittaker
Cartwright Hall
Lister Park
Bradford BD9 4NS (UK)
+44(0)274 493313

Chris Parsons Productions
123 Pembroke Road
Clifton, Bristol BS8 3EU (UK)
+44(0)272 237 922

Ravi Singh
Chrysalis Interactive Services
The Chrysalis Building
13 Bramley Road
London W10 6SP (UK)
+44(0)71 221 2213

Alex Morrison
Cognitive Applications
4 Sillwood Terrace
Brighton BN1 2LR (UK)
+44(0)273 821600

Paul Thompson
Design Museum
Butlers Wharf
28 Shad Thames
London SE1 2NJ (UK)
+44(0)71 403 6933

Alastair Layzell
Educational Television Network
PO Box 341
Jersey, Channel Islands (UK)
+44(0)534 43678

Dott. Angela Mottin
Hay Space Consulting Technologies
Via Legnano 6
20121 Milan (Italy)
Tel. +39(0)2 802 208415

Colin Pyrah
Heritage Projects
United House
Piccadilly
York Y01 1PQ (UK)
Tel. +44(0)904 646411

Steve Baker
I Contact
Century House South
North Station Road
Colchester, Essex CO1 1RE (UK)
Tel. +44(0)206 766099

Richard Land
IE
Studio 20
43 Carol Street
London NW1 0HT (UK)
+44(0)71 267 4923

Penny Ritchie Calder
Imperial War Museum
Lambeth Road
London SE1 6HZ (UK)
+44(0)71 416 5000

Fred Grainger
Interactive Learning Productions
Third Floor
North Street Court
North Street East
Newcastle-upon-Tyne NE1 8HD (UK)
+44(0)91 261 1255

Jonathan Carter
Jersey Museum Services
The Weighbridge
St. Helier, Jersey, Channel Islands
+44(0)534 30511

Neil Irwin
Light Fantastic Gallery
48 South Row
Covent Garden Market
London WC2E 8RF (UK)
+44(0)71-836 6423

Piers Lea
Line TV
99 Talbot Road
London W11 2AT (UK)
+44(0)71 792 8480

Rosalie Vicars-Harris
Media Projects International
7 Cameron House
12 Castlehaven Road
London NW1 8QW (UK)
+44(0)71 485 5657

Sarah Hennings
Moving Picture Company
25 Noel Street
London W1V 3RD (UK)
+44(0)71 434 3100

Paul George
Museum of Civilization
100 Laurier Street
POB Box 100, Station B
Hull, Québec J8X 4H2 (Canada)
+1 (819) 776 7007

Jonathan King
Museum of Mankind
6 Burlington Gardens
London W1X 2EX (UK)
+44(0)71-437 2224

Martin Oliver
National Museum of Photography, Film & Television
Prince's View
Bradford BD5 0TR (UK)
+44(0)274 727488

Des Fforde
National Waterways Museum
Llanthony Warehouse
Gloucester Docks
Gloucester (UK)
+44(0)452 25524

John Furlong
Natural History Museum
Cromwell Road
London SW7 5BD (UK)
+44(0)71 938 9157

Susi Barrett
New Media
Fredrica House, 12 Oval Road
London NW1 7DH (UK)
+44(0)71 482 5258

Tim Roderick
Norfolk TVEI
County INSET Centre
Wittard Road
Norwich NR7 9XD (UK)
+44(0)603 33276

Florence Cusson
Opus Species
14 rue de l'Atlas
Paris F-75019 (France)
+33(0)1 42 39 59 79

Chris Jennings
Research Machines plc
Hill Street
Oxford OX2 0BW (UK)
+44(0)865 791234

Dale Thompson
National Identity & Special Projects Task Force
Secretary of State's Department
15 Eddy Street
Hull, Québec K1A 0M5 (Canada)
+1(819) 997 3618

Richard Francis
Tate Gallery Liverpool
Albert Docks
Liverpool L3 4BB (UK)
+44(0)709 3223

Phil Pike
Tussauds Group
2 Silver Road
London W12 7SG (UK)
+44(0)81 740 4580

Mick Landmann
Vivid Interactive
Kingslake House
2-5 Union Street
Brighton BN1 1HA (UK)
+44(0)273 208730

Voyager Company
1351 Pacific Coast Highway
Santa Monica CA 90401 (USA)
+1 (213) 451 1383

Peter Howell
Xebec Multi-Media Solutions
Smith House
1-3 George Street
Nailsworth, Glos. GL6 0AG (UK)
+44(0)453 835482

THEMATIC INDEX

PLATFORMS INTRODUCED & DEFINED

DEFINITIONS & EXAMPLES OF OTHER TECHNICAL TERMS

DEVELOPMENT PRODUCTS

INSTITUTIONS & CLIENT COMPANIES

SUPPLIERS & DEVELOPERS

PUBLISHED DISCS